LOUISIANA

AS IT IS:

ITS TOPOGRAPHY AND MATERIAL RESOURCES—ITS COTTON, SUGAR CANE, RICE AND TOBACCO FIELDS—ITS CORN AND GRAIN LANDS—ITS NUMEROUS VARIETIES OF FIELD CROPS—ITS VALUABLE GRASSES—ITS FRUITS, INCLUDING THE ORANGE AND OTHER TROPICAL FRUITS—ITS VEGETABLE AND FLOWER GARDENS—ITS VAST AND VALUABLE FORESTS OF TIMBER—ITS PRAIRIES, BOTTOMS, SWAMP AND HILLY LANDS—HEALTH AND LONGEVITY—VARIOUS POPULAR ERRORS CORRECTED TOUCHING THE SOIL, CLIMATE AND PEOPLE OF THE STATE.

RELIABLE INFORMATION

FOR FARMERS, PATRONS OF HUSBANDRY, LABORING MEN, MANUFACTURERS, CAPITALISTS. MEN OF ENTERPRISE, INVALIDS—ANY WHO MAY DESIRE TO SETTLE OR PURCHASE LANDS IN

THE GULF STATES.

BY DAN'L DENNETT.

NEW ORLEANS
"EUREKA" PRESS, 33 NATCHEZ STREET.
1876.

The Object of this Work.

It is the object of this work to furnish facts in regard to the State of Louisiana, its material resources, health, local advantages, state of society, etc., that immigrants and strangers may rely upon with safety.

STATEMENTS STRICTLY RELIABLE.

The author of this book has been a resident of the Gulf States more than thirty-four years, and was editor and proprietor of the *Planters' Banner* most of the time from 1848 to 1872. It has been his object to publish nothing that cannot be vouched for by gentlemen of unquestionable intelligence and integrity.

In relation to the truthfulness of his general descriptions of Louisiana, and the usual fairness and reliability of his facts, published in his own and other journals, in regard to the material resources of this State and the inducements it holds out to immigrants from higher latitudes, he has permission to refer to:

A. M. Holbrook, Editor and Proprietor Picayune.
Dufour & Limet, Editors and Proprietors N. O. Bee.
J. N. Stoutemyer, Editor N. O. Times.
Editors N. O. Republican.
Louis J. Bright & Co., Editors and Proprietors N. O. Price Current.
Page M. Baker, N. O. Bulletin.
Thos G. Rapier, Editor Morning Star and Catholic Messenger.
J. Hassinger, German Gazette.
Linus Parker, Editor N. O. Christian Advocate.
Henry M. Smith, Editor S. W. Presbyterian.
Jas. H. Hummel, Editor and Proprietor Our Home Journal and Rural Southland.
E. K. Manard, Editor Son of the Soil.
Jno. B. Robertson, Editor Co-Operative News, and Author of Resources of Louisiana.
J. P. B. Wilmer, Bishop of Louisiana.
J. C. Keener, Bishop M. E. Church, South.
Rev. B. M. Palmer, D. D., LL. D.
Gen. G. T. Beauregard.
Charles Gayarre, Historian of Louisiana.
Charles J. Leeds, Mayor of New Orleans.
John Phelps, President N. O. Cotton Exchange.
Cyrus Bussey, President N. O. Chamber of Commerce.
W. R. Lyman, President N. O. Stock Exchange.

"LOUISIANA AS IT IS"

INDORSED BY THE LOUISIANA STATE GRANGE.

The State Grange of Louisiana (P. of H.) during its session in New Orleans, in December, 1875, indorsed, by a unanimous vote, the following resolutions offered in the Report of the Committee on Immigration:

1. *Resolved,* That we ask the favorable consideration of the State Grange, and of the members of our Order, to the forthcoming work of our Worthy Brother, Dan'l Dennett, so long characterized for his ability as a writer on the topography of Louisiana, and for his integrity of character as a Brother Patron. The title page and prospectus of his book, "LOUISIANA AS IT IS," is appended to these resolutions.

2. *Resolved,* That the State Grange of Louisiana cordially approves of the important publication of our Worthy Brother, Dan'l Dennett, on the rich and varied resources of this State, and will encourage the general circulation of his book, as the most reliable means of imparting correct information to immigrants, and to people of all sections and countries who desire to become acquainted with the hidden wealth of our State.

RICH'D M. McCORMICK,
Chairman.
RICH'D S. VENABLES,
Committee on Immigration.

WM. H. HARRIS,
Secretary Louisiana State Grange.

OFFICE OF THE STATE GRANGE OF LOUISIANA,
68 Camp Street, New Orleans,
April 24th, 1875.

We take great pleasure in stating that our Worthy Brother, Dan'l Dennett, for many years favorably known to the people of Louisiana as a gentleman of refinement and education, and as the editor of an agricultural newspaper, has for years devoted his talents to progressive agriculture and education in the South.

He is a Deputy of our Order in good standing, and being well acquainted with the geography, soil, and climate of Louisiana, is eminently qualified to give information to immigrants; and any statements he may make are entitled to full credence.

Executive Committee.
A. G. CARTER, Chairman.
THOS. RHINEHEART,
MICHAEL SCHLATRE,
H. W. L. LEWIS, Master La. State Grange.

WM. H. HARRIS, Secretary.

MATERIALS OF THE WORK.

We have gathered the materials which go to make up our book from various sources, always aiming to give full credit to the author of any article we may have appropriated in our columns. When we consider others' lamps, and lights, and oil, better than our own, we use them, when we have the owner's consent. Our object is to throw all the light we can possibly control on the natural wealth of Louisiana as it exists in 1876. Its political condition we have nothing to do with in this book; that is the work of statesmen and politicians.

We fill several pages from the pen of Col. S. H. Lockett, whose observations upon the topography of the State we highly prize. We republish a revised copy of our "DESCRIPTION OF THE ATTAKAPAS PARISHES AND ST. LANDRY." We published 20,000 copies of this circular, and a map of these parishes, in 1870. Many articles, and parts of articles, which have already appeared in different journals in Louisiana, will be reproduced in this book. It will be our object to use the marrow of what has been published on the material resources of the State, and add a large amount of new matter to complete the work.

The map that accompanies the book is furnished at a trifling cost to the purchaser,—a fine map of the same size would cost several times the amount paid for both book and map. By furnishing a cheap edition we are enabled to put 10,000 copies in circulation, otherwise we could have published but 5,000 copies.

From the N. O. Commercial Bulletin.

Information About Louisiana.—Daniel Dennett, Esq., a resident of this State since 1841, and honorably connected with journalism in Louisiana for more than twenty years; since the war devoted to the development of the resources of the State—a thorough, searching and painstaking agricultural and statistical reviewer, is about to publish a work calculated to do a great deal of good to Louisiana, in which he will give very reliable information interesting to farmers and manufacturers, to laboring men, to capitalists, to men of enterprise, and to all who desire to settle in the Gulf States or to invest their capital here.

The work of Mr. Dennett will be comprehensive, able, and in all respects trustworthy. It is a document which ought to be circulated all over Europe, and wheresoever else we expect or invite immigration to come from. It will contain a large amount of information interesting to every citizen, and especially to every officer and legislator. It is a document which ought to be embodied into our legislative proceedings, and translated into a half dozen languages; it should be scattered broadcast among the over-crowded people of Europe.

When the work appears, as it will shortly, we will speak more at length of its merits. We will only say now that it will supply a great desideratum. We do not hesitate to say in advance that the friends of Louisiana, and those who ardently desire the development of the resources of the Gulf States, will find in LOUISIANA AS IT IS a real treasure.

The Poetry of the History of Louisiana.

BY CHARLES GAYARRE.

The following brief extracts from Gayarre's excellent History of Louisiana, we trust will not be deemed inappropriate in this place, as an introduction to "LOUISIANA AS IT IS."

Under the head of "The Poetry of the History of Louisiana," after graceful and thoughtful preliminaries, the learned author remarks:

"I am willing to apply that criterion to Louisiana, considered both physically and historically; I am willing that my native State, which is but a fragment of what Louisiana formerly was, should stand or fall by that test, and do not fear to approach with her the seat of judgment. I am prepared to show that her history is full of poetry of the highest order, and of the most varied nature. I have studied the subject *con amore*, and with such reverential enthusiasm, and I may say with such filial piety, that it has grown upon my heart as well as upon my mind. To support the assertion that the history of Louisiana is eminently poetical, it will be sufficient to give you short graphical descriptions of those interesting events which constitute her annals. Bright gems they are, enriching her brow, diadem-like, and worthy of that star which has sprung from her forehead to enrich the American constellation in the firmament of liberty."

HERNANDO DE SOTO.

"On the 31st of May, 1539, the bay of Santa Spiritu, in Florida, presented a curious spectacle. Eleven vessels of quaint shape, bearing the broad banner of Spain, were moored close to the shore; one thousand men of infantry, and three hundred and fifty men of cavalry, fully equipped, were landing in proud array under the command of Hernando De Soto, one of the most illustrious companions of Pizarro in the conquest of Peru, and reported one of the best lancers in Spain.

* * * * * * * *

"Among his followers are gentlemen of the best blood of Spain, and of Portugal.

* * * * * * * *

"Now he is encamped in the territory of the Chicasaws, the most ferocious of the Indian tribes. And lucky was it that Soto was as prudent as he was brave, and slept equally prepared for the defence and for the attack. Hark! in the dead of a winter's night, when the cold wind of the north, in the month of January, 1541, was howling through the leafless trees, a simultaneous howl was heard, more hideous far than the voice of the tempest. The Indians rush impetuous with firebrands, and the thatched roofs which sheltered the Spaniards are soon on fire, threatening them with immediate destruction. The horses rearing and plunging in wild affright, and breaking loose from their ligaments; the undaunted Spaniards, half naked, struggling against the devouring element and the unsparing foe; the desperate deeds of valor executed by Soto and his companions; the deep-toned shouts of "St. Jago and Spain" to the rescue; the demon-like shrieks of the red warriors; the final overthrow of the Indians, the hot pursuit by the light of the flaming village, form a picture highly exciting to the imagination; and cold, indeed, must be he who does not take delight in the strange contrast of the heroic warfare of chivalry on one side, and of the untutored courage of man in his savage state on the other.

It would be too long to follow De Soto in his peregrinations during two years through part of Alabama, Mississippi and Tennessee. At last he stands on the banks of the Mississippi, near the spot where now flourishes the Egyptian named city of Memphis. He crosses the mighty river, and onward he goes, up to the White river, while roaming over the territory of Arkansas. Meeting with alternate hospitality and hostility on the part of the Indians, he arrives at the mouth of Red river, within the present limits of the State of Louisiana. There he was fated to close his adventurous career."

FATHER MARQUETTE AND JOLIET.

"One hundred and thirty years had passed away since the apparition of Soto on the soil of Louisiana, without any further attempt of the white race to penetrate into the fair region, when, on the 7th of July, 1673, a small band of Europeans and Canadians reached the Mississippi, which they had come to seek from the distant city of Quebec. That band had two leaders, Father Marquette, a monk, and Joliet, a merchant, the prototype of two great sources of power—religion and commerce—which, in the course of time, were destined to exercise such influence on the civilization of the Western territory, traversed by the mighty river which they had discovered. That humble monkish gown of Father Marquette concealed a hero's heart, and in the merchant's breast there dwelt a soul that would have disgraced no belted knight.

ROBERT CAVALIER DE LA SALLE.

"Seven years since the expedition of Marquette and Joliet had rolled by, when Robert Cavalier de la Salle, in the month of January, 1682, feasted his eyes with the far famed Mississippi. For his companions he had forty soldiers, three monks, and the Chevalier de Tonti.

* * * * * * * *

"Brought into contact with Count Fontenac, who was the Governor of Canada, he communicated to him his views and projects for the aggrandizement of France, and suggested to him the gigantic plan of connecting the St. Lawrence with the Mississippi by an uninterrupted chain of forts.

* * * * * * * *

"On the 15th of September, 1678, proud and erect with the consciousness of success, La Salle stood again in the walls of Quebec, and stimulated by the cheers of the whole population, he immediately entered into the execution of his projects. Four years after, 1682, he was at the mouth of the Mississippi, and in the name, (as appears by a notarial act still extant) of the *most puissant, most high, most invincible and victorious Prince*, Louis the Great, King of France, took possession of all the country which he had discovered.

* * * * * * * *

"To relate all of the heart thrilling adventures which occurred to La Salle during the four years which elapsed between the opening and conclusion of that expedition, would be to go beyond the limits which are allotted to me. Suffice it to say, that at this day, to overcome the one-hundredth part of the difficulties which he had to encounter, would immortalize a man.

IBERVILLE AND BIENVILLE.

"A few years after the death of La Salle, which occurred in 1687, by the hands of brutal companions, within the limits of Texas, a French ship of 42 guns, the Pelican, commanded by Iberville, after sinking an English vessel of 52 guns in a naval battle on the coast of New England, and whipping two other vessels of 42 guns each in the same fight, in the beginning of March, 1699, entered the Mississippi, accompanied by his brother, Bienville, and Father Anastase, the former companion of La Salle in his expedition down the river in 1682."

* * * * * * * *

In 1703 war had broken out between Great Britain, France, and Spain, and Iberville, a distinguished officer of the French navy, was engaged in expeditions that kept him away from the colony. It did not cease, however, to occupy his thoughts, and had become clothed, in his eye, with a sort of family interest. Louisiana was then left, for some time, to her scanty resources; but, weak as she was, she gave early proofs of that generous spirit which has since animated her.

* * * * * * * *

"Iberville sent his brother, Chateagué to the colony, accompanied by seventeen persons, as immigrants to the new colony.

"The excitement of this new arrival had hardly subsided when it was revived by the appearance of another ship, and it became intense when the inhabitants saw a procession of twenty females, with veiled faces, proceeding arm in arm, and two by two, to the house of the governor, who received them in state and provided them with suitable lodgings. But the next morning, which was Sunday, the mystery was cleared up by the officiating priest reading from the pulpit of the mass the following communication from the minister to Bienville: "His Majesty sends twenty girls, to be married to the Canadians, and to the other inhabitants of Mobile, in order to consolidate the colony. All these girls are industrious, and have received a pious and virtuous education. Beneficial results to the colony are expected from their teaching their useful attainments to the Indian females. In order that none be sent except those of known virtue and unspotted reputation, His Majesty did intrust the Bishop of Quebec with the mission of taking those girls from such establishments, as, from their very nature and character, would put them above all suspicions of corruption. You will take care to settle them in life as well as may be in your power, and to marry them to such men as are capable of providing them with a commodious home."

CONDENSED ITEMS FROM GAYARRE'S HISTORY OF LOUISIANA.

Sauvolle was the first governor of Louisiana. He died in 1701, and was succeeded by Bienville, the second governor.

Iberville went to France in 1701, and remained absent from Louisiana four years.

In 1706, the French girls brought to the colony were indignant at being fed on corn bread, and threatened to leave the colony on the first opportunity. This is called the "*petticoat insurrection.*"

Thirty-five colonists died of starvation in 1705.

The three most important personages in the Commonwealth of Louisiana at that time were Bienville, the governor, who wielded the sword, and was the great executive mover of all; La Salle, the intendant commissary of the crown, who had the command of the purse, and who, therefore, might be called the controlling power; and the Curate de la Vente, who was not satisfied with mere spiritual influence.

The commissary, La Salle, in a letter of the 7th of December,

1706, accused Iberville, Bienville, and Chateagué, the three brothers, of being guilty of every sort of malfeasances and dilapidations. The curate backed La Salle.

After an existence of nine years, the population of the colony did not exceed 279 persons. Its principal wealth consisted in 50 cows, 40 calves, 4 bulls, 8 oxen, 1,400 hogs, and 2,000 hens.

In 1709 famine re-appeared in the colony, and the inhabitants were reduced to live on acorns.

The scarcity of provisions had become such, that in 1710 Bienville informed his government that he had scattered the greatest part of his men among the Indians, upon whom he had quartered them for food.

In 1712, Anthony Crozat obtained from the King of France the exclusive privilege, for fifteen years, of trading in all that immense territory which, with its undefined limits, France claimed as her own under the name Louisiana. The charter of concessions virtually made Crozat the supreme lord and master of Louisiana.

In 1713 Cadillac is made governor of Louisiana.

The famous banking company of the Indies, with John Law at its head, was started in 1717. John Law was born in Edinburgh, Scotland, in 1671.

Bienville was appointed Governor of Louisiana the second time in 1718, and founded the city of New Orleans the same year.

In 1724 the white population of New Orleans amounted to 1,700 souls, and the black population to 3,300. In New Orleans there were about 1000 souls, including troops, and the persons employed by the government.

In the beginning of 1727, the spot where now stands New Orleans, not being protected by a levee, was subject to annual inundation, and presented no better aspect than that of a vast sink or sewer.

Mosquitoes buzzed, and enormous frogs croaked incessantly in concert with other indescribable sounds; tall reeds, and grass of every variety grew in the streets, and in the yards, so as to interrupt communication, and offered a safe retreat, and places of concealment to venomous reptiles, wild beasts, and malefactors, who, protected by these impenetrable jungles, committed with impunity all sorts of evil deeds.

In 1729 the French at Natchez were massacred by the Indians. The Indians captured and spared about 300 women and children whom they intended to make slaves of.

In 1733 the French colony in Louisiana was scourged by small pox and famine. A companion of Bienville wrote: "Our planters and mechanics are dying of hunger. The colony is on the eve of being depopulated. The colony is in such a state of indigence that, last year, the people were obliged, for more than three months, to live on the *seeds and grains of reeds.*

On the 15th of April, 1735, Bienville wrote, on the state of the colony: "One hundred thousand pounds of tobacco are made at Pointe Coupée; two women raise silk worms for amusement and succeed very well; eggs should be sent by the government to the Ursulines, who would teach this industry to the orphans, whose education is intrusted to them. The cultivation of cotton is advantageous, but the planters experience great difficulty in cleaning it from the seeds. Pitch and tar are made in some abundance."

In 1736 the colony turned its attention to the cultivation of indigo. But little was made from silk through ignorance of the business

The Balize Pass, which, in 1728, had sixteen feet of water; in 1738, fourteen feet and a half, and which Bienville represents as filling up rapidly, is known in our days as the Southeast Pass.

The Marquis of Vaudreuil was appointed governor of Louisiana in place of Bienville, in 1743.

De Vaudreuil, in 1744, commanded the planters to have their levees made, under the penalty of forfeiting their lands to the crown.

The Marquis of Vaudreuil marked the beginning of his administration by following the old and nefarious custom of granting monopolies.

In 1744 the white population of New Orleans was 800 souls, not including 500 soldiers, and the women and children. A few of the houses were brick, and the greater portion were wooden buildings, or were bricked up between posts. There were 25 inhabitants whose property was worth from one hundred thousand to three hundred thousand livres. Almost all the colonists were married. The most considerable of them was Mr. Dubreuil, who owned 500 negroes, several plantations, brick-kilns, and silk factories. At the German coast there were 100 white inhabitants, and 200 negroes. Occupations, gardening and grazing. Pointe Coupee, 200 whites and 400 negroes. Occupation, the cultivation of tobacco and the raising of provisions. Natchitoches, 60 whites, and 200 blacks. Productions, cattle, rice, corn, and tobacco.

The available force for defending Louisiana in 1746 was 400 white men, 500 to 600 Indians, and from 200 to 300 negroes.

A terrible hurricane in 1746, like that of 1740, destroyed the crops of the colony, and would have reduced the inhabitants nearly to starvation had it not been for boats from Illinois that annually supplied them with flour.

In 1751 the Jesuits sent some sugar canes from Hispaniola to the Jesuits of Louisiana, and some negroes who were used to the cultivation of this plant. The experiment was abortive, and though cane continued to be cultivated successfully, it was only in 1795 that the manufacture of sugar was successful.

On board the same ships which brought the first sugar cane, sixty girls were transported to Louisiana at the expense of the King. These girls were married to such soldiers as had distinguished themselves for good conduct, and who, in consideration of their marriage, were discharged from service. Such is the humble origin of many of our most respectable and wealthy families.

On the 23d of September, 1752, the Intendant Commissary, Michael de la Rouvillière, made a favorable report on the state of agriculture in Louisiana. "The cultivation of the wax tree, says he. "has succeeded admirably. Mr. Dubreuil alone has made six thousand pounds of wax. Some went to the seashore where the wax tree grows wild, in order to use it in its natural state. It is the only luminary used here by the inhabitants, and it is exported to other parts of America, and to France. In the last three years forty-five brick houses were erected in New Orleans, and several fine new plantations were established."

In 1753 Kerlerec took possession of the government of Louisiana.

The several governors of Louisiana paid flattering homage to the courage of the French colonists, and of the few Creoles or natives of Louisiana.

On the 3d of November, 1762, Louisiana was ceded to Spain.

In 1765, there was a considerable immigration to Louisiana from the Alibamons and Illinois districts, which had been ceded to the English, and from the province of Acadia, or Nova Scotia.

In 1755 the Acadian settlement at Grand Pré, Nova Scotia, was broken up by the English, under command of General Winslow, and the inhabitants, 1923 persons, were taken prisoners, and most of them transported to different States. Their houses and barns were burned by the English, and their property confiscated to the crown.

In 1765 about 650 Acadians had arrived at New Orleans, and from that town had been sent to form settlements in Attakapas and Opelousas under the command of Andry.

Ulloa, the new Spanish Governor, arrived in New Orleans on the 5th of March, 1766, and in February of the same year 216 Acadians arrived.

Ulloa at once ordered the census of the whole population of Louisiana to be taken, and the result was found to be: 1,893 men able to carry arms; 1,044 women, married and unmarried; 1,375 male, and 1,240 female children, Total, 5,562. The blacks were about as numerous as the whites. But the population was somewhat reduced by an epidemic closely resembling yellow fever.

A conspiracy was formed against the Spanish Government of Louisiana in 1768, and a general insurrection followed.

O'Reilly's administration, under the Spanish domination, continued one year, to 1770.

Unzaga's administration continued from 1770 to 1776; Galvez's 1777 to 1783; Miro's, 1784 to 1791; Carondelet's, 1792 to 1797; Gayoso's, 1797 to 1799; Casacalvo's, 1799 to 1801, Salcedo's, 1801 to 1803.

Louisiana was ceded to the United States in 1803-1804, for 50,000,000 francs, or $9,375,000.

Bonaparte, after the sale of Louisiana, said: "This accession of territory strengthens forever the power of the United States; and I have just given to England a maritime rival that will sooner or later humble her pride." * * * "The day may come when the cession of Louisiana to the United States shall render the Americans too powerful for the continent of Europe."

GOVERNORS FROM 1805 TO 1861.

Governor Claiborne—Governor James Villeré—Governor T. Bolling Robertson—Governor Henry Johnson—Governor Peter Derbigny—Governor A. B. Roman—Governor E. D. White—Governor A. B. Roman—Governor Alexander Mouton—Governor Isaac Johnson—Governor Joseph Walker—Governor P. O. Hebert—Governor Robert C. Wickliffe—Governor T. O. Moore.

CLOSE OF GAYARRE'S HISTORY.

LOUISIANA.

NATURAL DIVISIONS.

AREA.—Land and Water.—Arable Alluvial and Swamp Lands.—Good Uplands.—Prairie Regions.—Pine Hills.—Pine Flats.—Bluff Lands. Coast Marsh.

"Louisiana, though one of the earliest colonies of the continent, is as yet one of the few States concerning whose aspect and resources no reliable information is accessible, and whose climate and physical peculiarities are grossly misunderstood."—Prof. E. W. Hilgard.

AREA OF LOUISIANA.

The area of Louisiana has been variously estimated. Its numerous bays, lakes, bayous and rivers make it extremely difficult, if not impossible, to estimate its land and water surface with much accuracy.

Col. S. H. Lockett, lately and for many years Professor of Engineering in the Louisiana State University, who traveled over all the parishes of this State in search of data for his excellent topographical map of Louisiana, his various journeyings amounting to more than four thousand miles within the limits of the State, has taken much pains in working up a correct statement of the land and water surface of the State, and we prefer his figures and tables to any others we have seen. We give them below :

	SQ. MILES.
Lands within the limits of Louisiana, including coast marsh	40,790
Inland water and coast bays	2,328
Land and water surface	43,018

There are two general divisions of the lands of the State, the hilly and the level lands.

	ACRES.
Hilly lands	12,332,920
Level lands	13,773,000
Total	26,105,600

These lands may be classed as follows:

	ACRES.
Good Uplands	5,248,000
Pine hills	5,497,600
Bluff lands	1,587,320
Prairie region	2,483,000
Arable alluvial lands	3,615,000
Wooded alluvial	2,752,000
Pine flats	1,585,000
Coast marsh	3,338,000
Inland water surface	1,228,000
Coast bays	1,100,000

The sea-marsh, swamps, arable alluvial, and other divisions of land are so irregular in shape, and the boundaries of each division are so indefinite; there is so much surface that one cannot decide whether it is marsh or prairie, arable alluvial or swamp; it is so difficult to decide where the pine hills and the pine flats commence; and the flat country is so filled with lakes, bayous and marais of irregular shape, that no skill or labor can settle the question as to the number of acres in each division, or the number of acres of arable land in the State, or the water surface; but the facts which we have obtained are sufficiently accurate for all practical purposes, and Col. Lockett has taken more pains than ony other person in collecting these facts.

COL. S. H. LOCKETT'S DESCRIPTION.

The following interesting description of Louisiana was embodied in a lecture delivered by Col. Lockett before the Historical Society of New Orleans, in Lyceum Hall, in the winter of 1873. He has kindly permitted us to use the lecture in any manner that may promote the interests of the State.

A GENERAL DESCRIPTION OF LOUISIANA.

Louisiana, one of the Gulf States, lying between parallels 28° 50′ and 33° north latitude, is, with the exception of Florida and Texas, the most southern of the United States. It is included between the meridian 88° 40′ and 94° 10′ W. from Greenwich, measuring from north to south 200 miles, and from east to west, at its widest part, 290 miles, with a total area of 40,790 square miles, or 26,105,600 acres. It is bounded north by Arkansas, along parallel 33°, and by Mississippi along

31°; east by Mississippi, the Mississipyi and Pearl Rivers, forming the boundary line; south by the Gulf of Mexico; west by Texas, with Sabine River for the boundary line for more than two-thirds of the distance from Sabine Lake northward.

It is divided into fifty-six parishes, which correspond to the counties of other States. Arranged alphabetically, they are:

Ascension, Assumption, Avoyelles, Bienville, Bossier, Caddo, Calcasieu, Caldwell, Cameron, Carroll, Catahoula, Claiborne, Concordia, DeSoto, East Baton Rouge, East Feliciana, Franklin, Grant, Iberia, Iberville, Jackson, Jefferson, Lafayette, Lafourche, Livingston, Madison, Morehouse, Natchitoches, Orleans, Ouachita, Plaquemines, Pointe Coupee, Rapides, Richland, Red River, Sabine, St. Bernard, St. Charles, St. Helena, St. James, St. John Baptist, St. Landry, St. Martin, St. Mary, St. Tammany, Tangipahoa, Tensas, Terrebonne, Union, Vermilion, Vernon, Washington, West Baton Rouge, West Feliciana, Winn, Webster.

The total population of the State, according to the official census returns for 1870, is 726,276—of whom 362,651 are white and 364,210 colored. New Orleans is the capital and chief city.

ERRONEOUS VIEWS.

Few persons, not citizens of Louisiana, have a correct idea of its physical features.

As it is situated at the mouth of the greatest river on the continent, and contains within its limits the delta of that river, which all maps of the State show to be intersected by numberless lesser rivers and bayous, and filled with lakes, most people conclude that Louisiana is, throughout its entire extent, a low, wet, swampy region. They imagine its surface to be a great plain of wonderful fertility, where at all arable, with an indefinite succession of dense jungles, tangled swamps, marshes, lakes, sloughs, cane and cypress brakes. Such incorrect notions are apt to be formed by those who travel, within the limits of the State, along the Mississippi and other navigable rivers and bayous. But these misconceptions will be speedily dissipated by a journey into the interior. One will soon discover that few States of the Union possess a greater diversity of surface, soil, climate, scenery and products than Louisiana, and certain it is that no State has a more varied and interesting population, or a more wonderful history.

I will now endeavor to give the reader a good, clear idea of *Louisiana as it is,* as I myself *have seen it in my journeys through and over all of the parishes of the State, occupying more than two hundred days.*

NATURAL DIVISIONS AND SUB-DIVISIONS.

The whole surface of Louisiana may be divided into two grand areas, the hilly and the level country. The hilly parts may be subdivided into three regions which are characteristically different from one another in the configuration of the surface, in soil, in forest growth, and in fertility.

These subdivisions I have named respectively the good uplands, the pine hills, and the bluff lands. The level country may be subdivided into five classes, the pine flats, the prairies, the arable alluvial lands, the wooded swamps, and the coast marsh. Thus we have the whole land surface of the State subdivided into eight regions, which may be distinguished from one another by clearly marked differences.

There is still another large part of the surface of the State not included in the foregoing classification, which, in Louisiana especially, is of great extent and greater importance. I mean the water surface, which includes a vast multitude of rivers, bayous, creeks, lakes and bays.

By examining the map it will be seen that all of these subdivisions are plainly indicated after the manner of showing the different formations on a geological map. But there is a marked difference between my topographical map of Louisiana and the geological maps of other States. In the latter, the formations are distinguished from one another by the occurrence of certain lithological characteristics, or the presence or absence of certain fossils rarely seen and seldom known by the great mass of people. On the contrary, in the arrangement of the topographical subdivisions, the basis of classification has many circumstances with which every man in the State is perfectly familiar. The names of geological formations are either scientific, and therefore unintelligible to most persons, or purely local, and therefore not at all suggestive; while the names used to designate the topographical subdivisions are terms used in the common every-day language of the country, and possessed of a definite and well-established meaning. Nevertheless, the classification adopted in the topographical map is a true one, and I hope will have some interest for scientific men.

A PANORAMIC VIEW.

To get a good general view of the alluvial region, no place is better than the pilot house or hurricane deck of a Mississippi steamer when the river is well up. As the boat speeds along like a huge thing of life, a panorama of great beauty passes before the eye. The stately residences of the planters half hidden in groves of magnolia, pecan, and live-oak; the massive sugar-houses with their towering monumental chimnies, the neat villages of negro cabins whitewashed and systematically arranged, the broad fields of cotton, corn and cane, sweeping back in green waves to the blue line of distant woods, the fat sleek cattle and horses grazing on the levee, and verdant commons, are but some of the attractive features of the picture. This to the eye of a painter may not be comparable to the majestic mountain scenery of the Hudson, but to the agriculturist it would be difficult to present a more pleasing prospect.

A VIEW ON HORSEBACK, AND ON FOOT.

But along the banks of the great rivers we cannot see all that is worth seeing in the bottom lands. One must visit the deep, silent bayous, overhung with moss-covered cypress, willows and live-oaks; he must ramble along the clear, quiet lakes, whose polished surfaces reflect with perfect fidelity everything above and around them, save where float the broad leaves and bright flowers of the *graine-a-volet;* he must penetrate the tangled swamps, with their primeval forests standing as the representatives of past ages with their dense jungles of luxuriant cane, with their ponds, sloughs and marais, where the wild fowl rustles among the water-lilies; and if he has anything of an artist's eye, he will everywhere see new and peculiar beauties.

THE ALLUVIAL PARISHES.

My map shows that a great many parishes of the State contain alluvial lands within their borders. Carroll, Madison, Tensas, Concordia, Avoyelles, Pointe Coupee, West Baton Rouge, Iberville, Ascension, Assumption, St. James, St. John the Baptist and St. Charles, are mainly or wholly alluvia..

WEALTH AND REFINEMENT IN THE ALLUVIAL REGIONS.

To the population of this region the same remark applies as has been already made in regard to that of the bluff lands. The wealth, refinement, culture and hospitality of the Louisiana planter, regardless of his origin, whether it be French, Spanish, American, or of any other nationality, is too well known to need any mention from me.

NEGRO POPULATION.

There is one great and serious cause of trouble in the alluvial region since the late war. The negro population greatly outnumbers the whites, the disparity being very great, taking the average of the whole region. As a natural consequence, the negroes have become intensely interested in politics, to the great detriment of the crops and the agricultural interests of these parishes.

THE WOODED SWAMPS.

Of the wooded swamps little of interest can be said. They are scattered with great irregularity throughout the alluvial region, occurring in broader masses to the south than to the north of the mouth of Red River. A great part of the basin of the Atchafalaya is classed as wooded swamp, but in all probability much of it could be easily reclaimed. These swamps are of but little value to the State at present. Considerable quantities of cypress timber are felled in them and floated out in rafts for the use of saw mills along the Mississippi. No doubt a great deal of these swamp lands is reclaimable, but it is a matter of very little interest to Louisiana now when so much of her best land is untilled.

THE COAST MARSH.

The coast marsh borders the southern part of the State, from Pearl River to the Sabine. It extends from the water line of the Gulf of Mexico inland to a depth varying from ten to thirty miles, and averaging fifteen miles. It is low, wet, subject to tidal overflow, filled with lakes, intersected by numerous bayous, and generally impassable. It is, however, by no means uniform in its structure. At some points it is difficult to determine whether to designate the surface as a lake or as a grassy plain; at others, the marsh is firm enough for cattle to graze upon, and for the hunter to traverse; here, we shall encounter a "trembling prairie" whose upper surface looks firm and solid, while underneath a thin, treacherous crust, is an unknown depth of water and oozy mud; there, an island of live oaks growing upon rich, black, firm soil, capable of supporting the habitations of men.

ISLANDS IN THE MARSH.

Several of these islands are of considerable extent, and cultivated, such as Pecan Island, the Little and Grande Chenieres, Cheniere-au-Tigre, and the Buck Ridges of Cameron and Vermilion. They are all very fertile, producing cane, sea island cotton, oranges and bananas, have a delightful climate, and would be most desirable places of residence but for the swarms of mosquitoes which infest them, and the troublesome fly called "frappe-d'abord."

PARISHES WITH SEA MARSH.

The parishes which have the greater part of their surface covered with sea marsh are Cameron, Vermilion, St. Mary, Terrebonne, Lafourche, Jefferson, Plaquemines, St. Bernard and Orleans. In all of them there are other kinds of surface; belts of very fertile alluvial land along the bayous, some prairie in Cameron, a good deal in Vermilion and a less amount in St. Mary.

Excepting the planters living on the bayous, the population of the coast-marsh region is sparse, and consists mostly of hunters and fishermen. The products are cane and rice, oranges, bananas, figs, Japan plums, garden vegetables, fish and game for the New Orleans market.

WATER SURFACE OF LOUISIANA—LAKES.

The water surface of Louisiana, excluding the rivers and the bays, which open out into the Gulf of Mexico, is 1700 square miles. This includes Lake Pontchartrain and all the many fresh water lakes in the interior of the State. They are too numerous to justify a catalogue of them at this time; they are all shown on the map. Many of them are of considerable extent, and nearly all of them are beautiful sheets of water. They are an important element in the physical structure of the State, in the influence which they have on her climate, her commerce, and her agriculture, and are yet destined to be of still greater importance when Louisiana shall have been developed to that high state of prosperity for which she is so admirably adapted by nature.

HOMES ON THE LAKES—BEAUTIFUL BUILDING SITES.

No more beautiful sites can be found for summer resorts, for rural homes, or even for manufacturing purposes, than some of the lakes of Louisiana. They will all be noticed separately in my special descriptions of the various parishes in which they are found.

COAST BAYS AND LAKES.

If we include in the water surface of Louisiana those open bays and lakes on her coast, like Atchafalaya, Cote Blanche, Vermilion, Terrebonne, Timbalier and Barataria Bays, and Lake Borgne, the area will be about double the figures given above.

THE COAST LINE OF LOUISIANA.

Before leaving this part of our subject, it will be well to notice some peculiarities of the coast line of Louisiana. It may be divided into two distinct sections, differing from each other in many characteristic respects.

The first or eastern division lies between Cat Island, near the mouth of Pearl River and Atchafalaya Bayou, the southwest. These two points are the most easterly and most westerly limits respectively of the great delta of the Mississippi. The waters of the Mississippi formerly found their way through Manchac Bayou, Lake Maurepas, Lake Pontchartrain and the Rigolets into Lake Borgne, and thence into Mississippi Sound, at the entrance of which is Cat Island. These waters still flow into Atchafalaya Bay through the river of the same name. All this part of the coast is extremely irregular, indented with numerous bays, cut up by thousands of lakes and bayous into a labyrynth of peninsulas and islands, which it is almost impossible to represent on a map of the scale I have adopted. The general shape of this part of the coast is the arc of a circle, convex outwards. The radius of this circle is about sixty-five geographical miles, and its centre is a few miles to the westward of the southwest corner of Lake Pontchartrain. This circle crosses the narrow neck of land which makes the lower delta, near Forts Jackson and St. Philip. The whole length of the arc, excluding the lower delta, is one hundred and seventy miles. There is a remarkable tendency of the islands along this circle to form themselves into groups, convex towards the Gulf, and each island partakes of the same shape.

MARSH BAYOUS.

The bayous which flow through the marsh of this section are deep, except at their mouths, which are always obstructed by shallow bays with still shallower bars.

The second or western section of the coast from Vermilion Bay to Sabine Lake, is a nearly straight, continuous and regular beach, on the edge of the marsh. There are no outlying islands, and in fact the whole structure of the line is exactly the opposite of that of the other section.

THE GREAT PRAIRIE REGION.

From this comparatively uninteresting region we will now turn to what, to me at least, is the most pleasing part of the State—the great prairies. They lie almost entirely west of Bayou Teche and south of Bayou Cocodrie, making up the old Opelousas and Attakapas countries. On the south they are limited by the impassable sea marsh, into which they pass often by imperceptible gradation. On the west Calcasieu River and the Sabine form the boundary lines.

BAYOUS, COULEES AND FORESTS.

All of this extensive area, thus broadly defined, is not one unbroken, treeless expanse. Couleés and bayous course through it, generally in a north and south direction, on the borders of which grow fine forests of timber. From these principal belts of timber spurs run out into the open prairies, like headlands into the sea, thus dividing the whole region into separate tracts, each having its own name: Faquetyke, Mamou, Calcasieu, Sabine, Vermilion, Mermentau, Plaquemines, Opelousas and Grande Prairie are the largest. There are many others, with local names which it is needless to mention. The surface of the prairies, though generally level, is yet not perfectly so.

PRAIRIE BILLOWS, COVES AND HARBORS.

The prairie is gently rolling, like the billows of a deep sea. In fact, one cannot ride through the prairies without having their striking resem. blance to large bodies of water constantly recurring to his mind. The grass which grows upon their surface, waving in the wind, looks like ripples on the bosom of the ocean, the dark blue borders of woods are like distant shores, the projecting spurs like capes and promontories, the "coves" like bays and gulfs, and the occasional clumps of detached trees like islands in the sea.

SOIL OF THE PRAIRIES.

The soil of the prairies is either of a greyish yellow or a cold grey color, but is much better than is generally supposed, and improves wonderfully under proper cultivation. The subsoil is a good tenacious clay. The eastern part of the prairies has a better soil than that farther west. Yet, even the latter amply repays the laborer for his toil. By manuring, tramping, draining, and deep plowing, the prairie soil gets better every year that it is cultivated. Cotton, cane and rice may be raised with profit, and very probably, an excellent quality of tobacco. Hay, in any desired quantity, could undoubtedly be made, by enclosing parts of the prairie, and mowing the grass when fresh and juicy, or by improved and cultivated grasses.

PRODUCTS OF THE PRAIRIES.

The chief products of the prairies now are beef cattle and horses neither of very good quality, from the negligent manner in which they are raised. I saw a fine flock of sheep, and there is no reason why there should not be hundreds more. Poultry of all kinds can be raised with the greatest ease; vegetables and melons, figs, peaches and fine strawberries can all be grown successfully.

POPULATION OF THE PRAIRIES.

Most of the population of the prairies is of Acadian origin, and with but few notoble exceptions, they are not a thrifty people. They are kind, hospitable and sociable among themselves, but shy and suspicious of a stranger, especially if, like the writer, he speaks no French. Many of them are mere squatters on the prairies between the Vermilion river and the Sabine. Their houses, often half framed and half built of mud, are located sometimes on the open prairie, sometimes in the skirts of a belt of timber, and often without even a yard or garden inclosed. A neighboring marais will be surrounded by a rude "pieux" fence, and a small crop of rice raised. Their horses and cattle run at all times on the common prairie. *Cafe noir* is their nectar, and perique tobacco their ambrosia. With thousands of cows roaming on the prairies, you seldom see butter or milk in their houses. With the means around them of living well, they fare no better than the people who live on poor lands. Their educational advantages are poor, but they all learn to ride and to use a shot gun expertly about as soon as they learn to walk.

CLIMATE OF THE PRAIRIES—HEALTH AND LONGEVITY.

The climate of the prairies is admirable; breezy and cool in the summer, mild in the winter, dry and healthy at all times; the creole inhabitants are proverbially long lived. Altogether, this region may be regarded as naturally the loveliest part of Louisiana.

ALLUVIAL OR BOTTOM LANDS.

As an agricultural country the prairies cannot of course be compared with the alluvial or bottom lands, which we will now consider. These were once, and are destined to be again, the greatest source of wealth and prosperity to the State. They lie mainly along the course of the Mississippi. But an examination of the map will show that there are arms of considerable extent stretching out from the main body up the valleys of the Ouachita, Black, Little and Red Rivers. And like the tentacula of a huge polyp, there are narrow strips of arable alluvium following the courses of all the bayous that wind through the salt marsh seeking the Gulf of Mexico. All the smaller streams throughout the State have their narrow bottoms, but they are generally so variable in extent, now lying on one side of the parent stream, now on the other, and never more than a mile across, that I have made no attempt at tracing them.

GEOLOGICAL VIEW OF THE MISSISSIPPI BOTTOMS.

The manner in which the great bodies of the alluvium were formed is worthy of some attention, and will give us a better appreciation of its topography and its inestimable value. After the *bed* or foundation of the present "Mississippi Bottom" was washed out, and the bluffs already spoken of were formed, nature changed the plan of her operations. The bluffs resulted from a scouring-out, the alluvium from a filling-up process. Every tributary of the great river assisted in this work, while the old Mississippi itself was the chief architect.

The main stream rising in the beautiful Lake Itasca in the far northern part of Minnesota, comes down to this latitude with its waters laden with sediment drawn from the very oldest geological formations. The Ohio and its tributaries transport from Indiana, Ohio, New York, Pennsylvania, Virginia, Kentucky, Tennessee, North Carolina, Georgia and Alabama, the commingled materials of the Devonian and Carboniferous ages; the Missouri and Arkansas, with their hundred forks rushing from the slopes of the distant Rocky Mountains, and meandering through the vast plains of the West, bring in their turbid waters the materials of the cretaceous age, in addition to those of more recent formations. Red River, gushing two thousand miles away, from a deep cannon in the gypsum of the Ilano estacado, offers every year its fertilizing contribution; while all the lesser streams rob every hillside and mountain slope and smaller valley of their surface loam to pour it into the already overflowing lap of their common father. After gathering together all this rich material, the mighty Mississippi, inflated with pride, bursts his bounds and spreads his treasures with lavish prodigality over his "flood-plain," the child of his own creation. And thus he has been renewing and enriching this favored child for ages. No wonder then if the alluvial lands of Louisiana are the most fertile in the world; no wonder, with so composite a nature, if they are of inexhaustible strength; no wonder, thus piled up year after year, century after century, if they are deep beyond any parallel!

This, the manner of their formation, determines the topography of the alluvial lands. They are always highest immediately on the banks of the streams, and slope off into low and densely wooded swamps. Every overflow makes the high ground higher, and fills up part of the swamps, diminishing both their depth and extent,

WONDERFUL FERTILITY OF THE ALLUVIAL LANDS.

Throughout the entire length of the alluvial region, wherever the soil is arable at all, it produces in perfection all crops suitable to the latitude. Cotton, corn, cane, rice, tobacco, with the juicy orange and delicious banana, all combine to make the Louisiana planter the prince of agriculturists. In nearly every part of the alluvial region there are found two distinct varieties of soil.

TWO VARIETIES OF BOTTOM LAND SOIL.

The lands nearest to the streams, called front lands by the planters, are moderately sandy and easily worked, The back lands, on the contrary, are stiff, sticky, and difficult to break up. Both classes are very fertile, but for most purposes, and especially for sugar-cane, the front lands are preferable. The color of these lands very plainly indicates their origin.

RED RIVER AND MISSISSIPPI SOIL DEPOSITS.

All direct formations of the Mississippi are dark colored and often nearly black; those of Red River and its confluents, Bayou Bœuf, Bayou Teche, etc., are of a warm light-red tinge, or of a deep Indian-red color. Along the Atchafalaya and the net-work of streams connected with it, the dark soil predominates in some places, the reddish in others, showing that the overflows of that river are caused sometimes by Red River, sometimes by the Mississippi.

BLUFF LANDS.

We now come to the bluff lands, one of the most peculiar and interesting features of Louisiana. It is difficult to give a clear description of the bluff region, and an intelligible idea of its topography, without first considering its geological origin. I am not a professor of geology, or a skilled geologist; not at all conversant with all the multifarious details of that important and complex science. But no one can thoroughly study the topography of a country without understanding the general principles of geology, and I hope I may claim to possess such a knowledge, and may make bold to give in simple language my notions concerning the origin of the bluff regions.

GEOLOGICAL VIEW OF THE BLUFF LANDS.

The good uplands and pine hills were formed during periods of great continental commotion. They belong to the glacial and drift epochs of the geologists, during which all-pervading currents of water were poured with mighty rush from north to south, bringing down with them the immense volumes of sand and pebbles with which all our hill country is covered.

After this came a period of comparative repose; the universal north and south currents ceased to flow, and the waters of the old Mississippi River found their way to the Gulf through a broad, placid and comparatively shallow estuary, which extended inland to the junction of the Ohio and Mississippi, with arms stretching out up the valleys of the other tributary streams.

During this reign of quiet, fine mud and silt were washed into the estuary from the higher lands, and floating gently towards the sea, were deposited in a deep stratum, whose upper surface was a broad subaqueous plain. Thts stratum of silt completely filled up what we now call the Mississippi bottom, and the broad plain extended out to the hillside slopes on the east and west to a distance of sometimes eighteen or twenty miles from the limits of the present alluvial lands.

The city of Baton Rouge and all the bluff cities on the banks of the Mississippi stand upon that once submerged plain. But it is now no longer under the dominion of the waters, for a subsequent and the last great continental movement took place, and this plain was raised several hundred feet above its old-time level. The old Mississippi had then to wash out for itself a new channel to the sea, and having the fine, easily dissolved silt, which had not been indurated by exposure to the sun, to work upon, the mighty river swept much of its former bed into the Gulf of Mexico. In so doing there were left on either side those high, steep bluffs which form so striking a characteristic of its shores. On the east, Memphis, Vicksburg, Natchez, Port Hudson and Baton Rouge stand on these bluffs, all immediately on the banks of the river. Back of all these cities, from ten to twenty miles, extends the once submerged plain, now an elevated plateau, or a succession of high hills, which is indicated on the map as part of the bluff region. On the west, none of the bluffs, at least within the limits of Louisiana, are now very near the river. Nevertheless, they can be easily traced.

LOCALITY OF THE BLUFF LANDS.

The Bayou Maçon hills in Carroll parish are the beginning of these bluffs in the north. We follow these hills through Carroll, Richland and Franklin to the mouth of Deer Creek; thence we trace the bluffs in Sicily Island; thence along the base of the pine hills of Catahoula and Rapides to Grimes' Bluff on Red River in Avoyelles; thence through Avoyelles prairie and the coteaux south of it to Bayou Rouge prairie; thence we skip across a belt of alluvium to Moundville on Bayou Bœuf, and continue southward through Washington, Opelousas and the Grand Coteau of St. Landry; thence we follow the bluffs along the Carrancros Hills and the hills of the Cote Gelee; thence finally through those fine lovely islands which stud the dreary salt marsh like the fabled jewel in the head of the loathsome toad. Back of this line, thus traced at rapid strides, lies another bluff region.

GENERAL APPEARANCE OF THE BLUFF LANDS.

The topography of this region results directly from its geological origin. We find it everywhere either a level plain with sinuous gorge-like ravines intersecting it, or else it has been washed into the most strangely irregular ridges and hills that any one ever encountered. Vicksburg, a well known locality in Mississippi. is a representative of the hilly bluff. Port Hudson and Natchez are types of the level bluft with the deep ravines. The Tunica hills of this State are like those around Vicksburg, Sicily Island, Avoyelles Prairie, Opelousas, and in fact all of the western bluff regions, except the five islands, are more like those in the vicinity of Port Hudson. The Bayou Maçon hill country belongs to the level type, but it is not so elevated as other similar regions. I account for this fact by supposing that the great volumes of water brought down from the mountains of the West by the numerous streams which flow from that direction, have washed away the upper strata, and left on the surface what corresponds to a very low stratum in the Port Hudson bluff. This may also account for the fact that the Bayou Maçon hills are not quite so fertile as other bluff regions.

CHARACTER OF THE SOIL—FOREST GROWTH.

The soil of the bluff lands, generally, is very fertile. It is a fine calcareo-silicious silt, of a yellowish grey color, easily cultivated, washes fearfully, gets muddy with the least rain, and becomes intolerably dusty in very dry weather. The forest—always magnificent—is made up of oaks of all kinds, especially the white, and the water varieties, magnolias, beeches, poplars, hollies, dense cane-brakes, tangled masses of muscadine, fox-grape and other vines. Depending upon latitude, the bluff lands produce, with great success, all the staples of Louisiana. Scattered through them are open spots with very infertile soil, and shallow, marshy ponds, with a fringe of rank grass called marais. In the wooded portions the marais are replaced by low wet places called spicewood swamps, or slashes. The great prairies belong, geologically, to the bluff formation, but topographically they are so different, I have given them a separate color on the map and a distinctive description.

THE FIVE ISLANDS OF ATTAKAPAS—THE SALT MINE.

The most interesting of all the bluff lands are the five islands, Miller's, (or Orange), Petite Anse, Cote Blanche, Grande Cote, and Belle Isle. They are really wonderful to behold, looming up on the great marsh like mountains in the sea. The highest of them is nearly two hundred feet above the surrounding plain, and the largest is not more than two miles across. Everything about them is striking and peculiar. Petite Anse is the most notable, but they are all worthy of a visit. That marvelous salt mine on Petite Anse can be likened to nothing so fitly as to the palace built by the magic of Aladdin's lamp. The wonder is, how did it get there, and by what mysterious agency was created so great a mass of pure rock salt, whose galleries, excavated by the miner, glisten and glitter like halls of pure crystal studded with diamonds!

But I must not delay the reader longer on the bluff region. I have not exhausted the subject, but we are not half through with Louisiana yet. The parishes in whieh the bluff lands occur have been pretty well indicated in tracing the boundary lines of the formation. It is needless to say anything of the inhabitants. As these lands are scattered from the extreme northern boundary of the State to the Gulf coast, there is no classifying their population. Many of the wealthiest, most refined and intelligent people in Louisiana live on the bluff lands.

THE PINE FLATS.

I will now notice the pine flats. Of these little can be said. By referring to the map, the reader will see that two widely separated areas are thus designated. In the eastern part we find the pine flats occupying the southern half of St. Tammany, not quite so much of Tangipahoa, and nearly half of Livingston. In the extreme southwestern part of the State is an area of about the same extent, in Calcasieu parish.

SOIL OF THE PINE FLATS.

The soil and forsest of the flats are very similar to those of the pine hills, the soil being sandy, thin and poor, the forest the long-leaf pine. But the surface is almost a perfect level. They contain many wet, oozy spots,, called "bay galls," from the clumps of bay trees always found in them, and here and there will be seen a shallow, marshy pond, or marais, similar to those of the bluff lands and prairies.

OPEN WOODS AND PINE FOREST PASTURES.

The woods are so open that, traveling along the lonesome roads through the flats, the rider may often see a herd of deer quietly browsing a half-mile away from him. The surface is covered with grass, and makes a good range for cattle. A great deal of lumber, both pine and cypress, is cut, the latter occurring in abundance on the banks of the larger streams, which flow through the pine flats. Turpentine might be manufactured in unlimited quantity, but this industry has never been developed in Louisiana.

INHABITANTS OF NORTH LOUISIANA.

The inhabitants of North Louisiana are different in many respects from those of the low country. They are generally of English, Scotch or Irish descent, immigrants from the older and more eastern States, or the offspring of such immigrants. Georgia, Alabama, the two Carolinas and Virginia, have all furnished their quota towards making up the population of this part of Louisiana. They are a thrifty, energetic, enterprising people. Their soil is not so exuberantly fertile as that of the low country; they know they must work it to make a living from it, and they do so with a will, and with commendable perseverance.

VILLAGES IN NORTH LOUISIANA.

In South Louisiana we see immense plantations and few towns; in North Louisiana much greater numbers of small farms, and numerous thriving villages, each with its two or three churches, "male" and "female" schools, masonic lodges, and perhaps a factory of some kind on a small scale. North Louisiana is pretty generally, but not very thickly settled. There is room still for thousands of families, who can readily obtain very fair lands at very reasonable prices. The climate is healthy, except in the immediate vicinity of creek and river bottoms, which are subject to overflow. The water throughout the good uplands is abundant, and of excellent quality. Good springs are common, and sweet free-stone water may be had almost anywhere by digging a well from twenty to sixty feet deep.

REGION OF THE PINE HILLS.

The next subdivision of our map is the region of the pine hills. It takes but a few words to describe it. The surface is broken and hilly, the elevation of the main ridges but little, if any, less than that of the good uplands, The soil is generally poor and sandy; ledges of course gray sand-rock are very common, and immense numbers of petrifactions. The forest growth, almost to the exclusion of every other tree, is the long-leaf pine, and the scrub black-jack oak. Where the Red River and other larger streams of the State flow through the pine hills, of course we find the usual bodies of fine alluvial lands along them, and on all the lesser streams are narrow bottoms which afford room for small but tolerably good farms.

HOG-WALLOW LANDS.

There are in the pine hills some peculiar tracts, called by the inhabitants hog-wallow lands. They are characterized by a stiff, sticky, calcareous soil, which becomes terrifically muddy in wet weather. The forest growth upon them is principally the post oak and black-jack, intermingled with the pines. The soil is not at all fertile, and has nothing to recommend it. These hog-wallow lands extend in an almost continuous belt across the region of the pine hills from the mouth of Anacoco Bayou; in Vernon, to Grand View, on the Ouachita. In this belt are several small open or "bald" prairies.

WOODS, PASTURES, STREAMS OF PURE WATER.

The piny woods are quite open, with but little underbrush. The surface is covered with a coarse grass which makes a pretty fair pasturage in the spring, Throughout the whole region there are numerous clear, bold streams of pure water, all of them abounding in fish, and many of them affording fine mill sites and water power sufficient for factory purposes.

VAST SUPPLIES OF PINE LUMBER.

The pine hills make some good upland cotton; corn and potatoes are raised for home consumption; some beef cattle are sent to the market; and they could supply the world with pine lumber of the very best quality, and might produce turpentine in any desired quantity.

PINY WOODS INHABITANTS.

The inhabitants of the piny woods are proverbially poor, but honest, moral, virtuous, simple-hearted and hospitable. In some neighborhoods that I passed through I found little communities banded together by kindship or long friendship, with many of the evidences of thrift, comfort and prosperity around them.

BUNDICK'S CREEK SETTLEMENT.

On Bundick's Creek, for instance, in Calcasieu, I stopped for several days in such a settlement, where I found a church and schoolhouse, each family with its herd of cattle, flock of sheep and drove of hogs. They all made their own cotton, corn, potatoes, rice, sugar and tobacco; spun and wove their own clothes for summer and winter wear, made their own leather, and in fact, supplied nearly all of their necessities by home products, with the exception of coffee and a few articles of dry goods. And although their cotton crops had to be hauled to find a market, either to Alexandria or Lake Charles, a distance of over half a hundred miles, yet every head of a household had a goodly roll of "greenbacks" put away in the chinks of his log cabin. There are hundreds of settlements in the pine hills, which, with the same management,, might be quite as well off as the one just noticed.

PINY WOODS PARISHES.

The piny woods parishes are Catahoula, Winn, Grant, Natchitoches, Rapides, Vernon and Calcasieu, west of the Mississippi; St. Helena, Tangipahoa, Washington, St. Tammany and Livingston, east of that stream. They all have other kinds of land in them besides that covered by the long-leaf pines. Catahoula has a splendid body of bluff land in Sicily Island, and a considerable tract of alluvium on the borders of the Ouachita and Black Rivers. In Natchitoches there is the beautiful Cane River country; in Grant are the alluvial lands between Bayou Rigolet and Red River; in Rapides, the unsurpassed lands of Bayous Rapides, Robert, and Bœuf. In Calcasieu there is a large area of prairie, and in Washington and St. Tammany a strip of alluvium along Pearl River.

TOPOGRAPHICAL SUBDIVISIONS.

I will now take up the topographical subdivisions separately, and in the order in which they haye been named, and describe them somewhat fully, believing that we shall get a more perfect understanding of Louisiana as a whole by first thoroughly understanding its component parts.

GOOD UPLANDS.

The good uplands lie mostly in North Louisiana, and cover the greater part of the parishes of Caddo, DeSoto, Sabine, Bossier, Webster, Red River, Claiborne, Bienville, Union, Jackson, Ouachita, Morehouse and a small part of Caldwell. In the "Florida parishes," east of the Mississippi, a tract of country similar to the good uplands of North Louisiana, is found in the eastern part of the two Felicianas, and in the northeastern corner of East Baton Rouge.

HILLS AND VALLEYS—ELEVATION.

A great part of the good uplands region is extremely hilly, some of the ridges being almost mountainous in elevation, reaching a height of at least three hundred feet above the valleys of the main streams between them, and from four to five hundred feet above the Gulf of Mexico. The elevation of this region I gather in the following manner: By a section line made in the surveys of the railroad from Monroe to Shreveport, it appears that one point crossed in Jackson parish is 250 feet above the reference point on the banks of the Ouachita opposite Monroe. According to the report of Humphreys and Abbott, this reference point is 12 feet below the banks of the Mississippi opposite Vicksburg, which latter is 100 feet above the Gulf. From these data we calculate the height of our first point to be 338 feet above the Gulf. Still farther westward, near Arcadia in Bienville, the proposed line of railroad reaches a summit of 299 feet above the reference point, and therefore 387 above the Gulf. Some of the trial lines, I was informed, passed over parts of the ridges of even greater elevation, and as it is not all likely that a line of survey would have passed over the highest points of any ridges to be crossed, we may safely estimate the elevation of the loftiest summits in this part of the State at over 400 feet above the Gulf of Mexico. Farther north the ridges are still higher, so that we would surely be within the limits of safety in assuming 500 feet as their probable maximum height.

CHARACTER OF THE SOIL.

Although the surface of the good uplands region is so elevated, and frequently quite broken and rough, yet the soil is generally pretty fair. Commonly it is gray sandy, easily worked, and rich enough when freshly cleared, but soon washes away under the present system of careless cultivation. The subsoil of the sandy country is a hard and sandy clay, which is always exposed in the old fields and roads, and gives to the vicinage of all the villages and older settlements of North Louisiana a very worn and ragged appearance.

There are notable exceptions to the gray sandy soil in the good uplands. In Morehouse, Ouachita and Caldwell on the east, and Caddo, DeSoto and Bossier on the west, there are large areas covered with a mellow, yellow, loamy soil of very considerable fertility. And in Sabine, Webster, Claiborne, Union and Jackson are many belts of land of several miles in length, and a mile or more in width, with a strange blood-red soil, from a strong impregnation of iron, that are very rich. These belts are aptly named the "red lands." They are generally found on the summits of the highest ridges, are covered with rough, angular fragments of ferruginous sandstone, or iron rock, and look exceedingly unpromising to a stranger. But their external features are no indication of their fertility and durability. At Greensboro, in Jackson, I saw a farm on the red lands which had been cultivated for forty years, and yet made forty bushels of corn to the acre, and three-quarters of a bale of cotton, in spite of the fact that no care had been taken to preserve it.

FOREST GROWTH OF THE GOOD UPLANDS.

The prevailing forest growth of the good uplands is a very good one of mixed timber. Oaks of various kinds, but principally of the red, white, black and post oak varieties; the dog wood, beech, sassafras, hickory, black gum, sweet gum, ash, maple, and the shortleaf pine, constitute the larger growth of bushes, such as the hackberry, chinquapin, elder, sour wood, prickly ash, etc., with many fox grape and muscadine vines.

BEST COTTON LANDS IN THE WORLD.

All of the parishes above named have tracts of greater or less extent of first class alluvial lands lying along the various streams which flow through them. DeSoto, Bossier, Caddo and Red River parishes have some of the best cotton lands in the world on the banks of Red River, Sabine, Jackson, Claiborne and Bienville have no extensive areas of alluvial lands; yet every creek, bayou and rivulet within their limits—and there are many of them in each—has its first and second bottoms of excellent land. In Morehouse, Ouachita, Union and Caldwell there are some exceedingly fertile lands on either side of Ouachita River. Bienville, Jackson, Ouachita, Caldwell and Sabine have considerable areas of the long-leaf pine country, of greatly inferier fertility to the rest of the good uplands. It is to be understood in every instance where a parish is mentioned as belonging to a certain subdivision, that the larger part of its surface is comprehended in that subdivision.

SCENERY OF NORTH LOUISIANA.

North Louisiana is not altogether devoid of attractive scenery. From the tops of the lofty red land hills there are some wide and pleasing prospects. Such scenes, however, are pleasing only from their vast extent. But there are some beautiful and really picturesque landscapes along the meanderings of the Ouachita, on Lakes Bodeau and Bisteneau, in the vicinity of the famous Red River Raft, and among the numerous lakes scattered all along the Valley of Red River.

PRODUCTS OF THE GOOD UPLANDS.

The products of the good uplands are cotton, corn, sweet potatoes, and small grain to some extent, also, sugar cane and rice do very well in these parishes. The red lands especially are well adapted to the culture of wheat and rye. In North Louisiana, peaches, pears, apples and figs all do well, and no country is better suited to the cultivation of melons and all the most useful garden vegetables. Grape culture has also been tried with considerable success.

This part of Louisiana is very similar to a large part of all the States lying east of it to the Atlantic coast. It has its exact counterpart in the central parts of Mississippi, Alabama and Georgia. I suppose the same belt extends westward into Texas.

VAST NATURAL WEALTH OF LOUISIANA.

Thus far, everything we have said about Louisiana is all *couleur de rose*, and yet our brief description of the State has been by no means too highly colored. In fact when we consider her natural resources, and the advantages she possesses from her climate, soil and location, it is simply impossible to speak of her in terms of praise too strong. But there must be shade and shadow in every picture, and so there is in this, and we must look on the dark as well as on the bright side. All that nature has done for Louisiana will amount to nothing, unless man does his part by her. The wealth stored up in her inexhaustible soil, though it be as great as that of Ophir, or that once hidden in the quartz rock and sand of California, will remain forever buried, unless brought to light by intelligent and persistent labor. And so of her many other resources, which are not directly dependent upon the fertility of her soil, they will not spontaneously develop themselves.

DUTIES NEGLECTED—THE ROAD TO SUCCESS.

I need but mention a few industries not now in activity to show how much requires to be done and might be done, to bring the State up to her capacity for wealth and prosperity.

Avery's salt mine, not more than a hundred miles from New Orleans, with easy transportation to that city, sends not a pound of salt to the market.

The most wonderful deposit of sulphur in the world exists in Calcasieu parish, has been known for several years, and yet brimstone is imported into Louisiana for the use of the sugar mills.

With an everlasting carpet of sweet grass an the prairies, which would make excellent hay if mowed and cured early enough in the season, hundreds of thousands of bales of long forage are annually brought to Louisiana from other States.

With the richest soil in the world, not a boat comes down the Mississippi from the Northwest but is laden to her guards with corn, flour, potatoes, cabbages, and other provisions.

With good water-power everywhere in the hills, and water enough in the low lands to run all the engines in the world, there are no factories worth mentioning.

With level fields and open prairies admirably adapted to machine cultivators, I have never yet seen one in the State.

With so many thousands of square miles of pine timber, not a barrel of tar, pitch or turpentine is made for commercial purposes.

With forests of timber, trees adapted to every kind of wooden fabric, from the most delicate household furniture to the line-of-battle ship, everything made of wood—even to ax-helves—is imported from St. Louis, Cincinnati, and New England. Nay, more, hundreds of thousands of the best cotton and cane lands in the State, once cleared and cultivated, are now lying idle, abandoned and growing up in cockle weeds, coco grass and blackberry vines. I might go on thus and make a score more of similar statements, but these are enough to show the drift of my argument. Intelligent, earnest, persistent labor is needed everywhere throughout the State. There is more scope for energy and enterprise in Louisiana than in any other State in the Union; there are more opportunities for the judicious investment of capital; there is more room for honest, industrious, laboring men.

The abandoned lands alone will accommodate at least twenty thousand families of farmers; the prairies could feed the flocks and herds of nearly as many more of those who would prefer stock raising; while all through the bluff lands, the good uplands, and the pine hills, good homes on good soil in a healthy climate may be found in almost every section of land by such as are willing to work.

IMMIGRATION NEEDED.

Let the people of Louisiana appreciate the true condition of their State, understand fully her true needs ; let them see that she wants tens of thousands of honest, energetic, industrious *white men* to immigrate to her waste places.

Let them adopt a broad, liberal, and practicable scheme for encouraging such immigration ; let them spread broadcast over the more thickly settled States of this country and Europe invitations to immigrants to come to her.

Let them show to all who would come that Louisiana is abundantly able to maintain them, and a new and brighter era will have dawned for her, and soon she will not only be restored to her former state of wealth and prosperity, but will eclipse by her future splendor all that she has been in the past.

SOUTHWESTERN LOUISIANA.

BY D. DENNETT.

A Description of the Parishes of St. Landry, Lafayette, St. Martin, Iberia, Vermilion, and St. Mary.

In which will be found, in brief paragraphs, statements descriptive of these Parishes; their Geography, Topography, Climate, Soil, Prairies, Forests, Swamps, Marshes, Lakes, Bays, Bayous, Islands, Scenery, Productions, Crops, Fruits; Yield per acre of the Staple Crops: Domestic Animals and Fowls; Prairie Herds; Wild Game; Fishes; Navigation; Health, Mineral and other Resources; Markets; Railroads; Schools; Churches; Prices of Freight and Passage; Aggregate Sugar Crops; and other matters of Interest to Strangers and Immigrants.

GENERAL DESCRIPTION.

Seventy-three miles west of the city of New Orleans, the Morgan Louisiana and Texas Railroad crosses the Bayou Bœuf, the eastern boundary of the Parish of St. Mary; and seven miles further west is Brashear City, on Berwick's Bay, the western terminus of all of this railroad which is now completed.

About 110 miles west of Berwick's Bay is the mouth of the river Mermentau, which receives the waters of the Nez Pique, through the upper Mermentau, Lake Arthur and Lake Mermentau.

From the northern boundary of St. Landry to the Gulf Coast the distance is about 100 miles; and from Belle River, the eastern line of the Parish of Iberia, to Lake Arthur, the western limits of the Parish of Vermilion, the distance is about eighty-five miles. The sea marsh on the coast of Attakapas has an average width of more than twenty miles. Its greatest width is about twenty-seven miles.

The southern boundary of these parishes is in latitude twenty-nine and a half degrees—almost half a degree south of the latitude of New Orleans. The northern limits of St. Landry reach latitude thirty-one, near the true cotton belt of the Southern States.

The five parishes, St. Mary, Iberia, Vermilion, St. Martin and Lafayette, were originally called ATTAKAPAS, and they are now called "The Attakapas Parishes." The name, Attakapas, comes from the name of a tribe of Indians that once inhabited this country.

NAVIGATION AND TRAVELING FACILITIES.

The passenger train on the Morgan Louisiana and Texas Railroad arrives at Brashear City, from Algiers, opposite New Orleans, daily at 12 o'clock M., and departs for New Orleans at 1 o'clock P. M.

The Morgan Texas steamships, the finest on Southern waters, run daily between Brashear and Galveston.

The steamers of the Attakapas Mail Transportation Company leave Brashear City daily, for New Iberia, a distance of 72 miles, halting at Pattersonville, Centreville, Franklin, Charenton and Jeannerette, and at intermediate landings. They usually extend their trips to St. Martinsville three times a week, 102 miles from Brashear.

Steamers, in 1870, frequently made trips to Avery's Salt Mine, on Petit Anse Bayou, for cargoes of salt and sugar. The trips between Brashear City and the mines have been made in ten hours, a distance of about 75 miles.

A steamer sometimes makes trips between St. Martinsville and New Orleans via the Atchafalaya, the mouth of Red River and the Mississippi. It takes about nine days to make the round trip.

Steamers leave Washington twice a week for New Orleans via the Courtableau, Atchafalaya, the mouth of Red River and the Mississippi

Frequently half a dozen or more small jobbing boats are employed on the Attakapas waters, some of them making trips to Vermilion river, Grand Cote, Cote Blanche, Belle Isle and the mouth of Bayou Sale. They do a large business towing rafts of cypress logs for the saw mills, from the bayous across Grand Lake, and in bringing pieux and other split lumber from the lakes to the planters of the Teche.

United States Mail coaches leave New Iberia three times a week for Washington, Parish of St. Landry, passing through Vermilionville, Grand Coteau and Opelousas, and tri-weekly U. S. Mail coaches leave Washington for Alexandria.

A coach runs with considerable regularity between New Iberia and St. Martinsville, a distance of nine miles by land.

A horseback mail extends to all the postoffices of these parishes off the main traveled routes. A mail coach runs regularly between New Iberia and Abbeville.

TILLABLE LAND.

These six parishes contain more than three million acres of tillable land, most of it of inexhaustible fertility. Even most of the sea marsh, and all of the swamp lands, may be reclaimed by local levees and drain ing machines, and may become the most productive rice and sugar lands in the State. Windmill pumps may relieve the relaimed marsh-lands from surplus water, for the winds blow almost constantly near the Gulf Coast.

DIVIDING LINE.

On the border of the sea marsh of St. Mary and Iberia, extending from a point below Berwick's Bay to and into the Parish of Vermilion, a line of forest trees, mostly heavy cypress, stand as the dividing line and wall between the marsh and the tillable lands of the Atchafalaya and the Teche. In places this line of timber is from one to two miles wide, and even wider. This line of forest extends down to the mouth of Bayou Salé, on both sides, and down both sides of Bayou Cypremont. At Petit Anse Island the sea marsh and prairie meet, and the chain of timber is broken for a few miles. On the side of this crooked chain of timber, next to the plantations, in places, there is a heavy growth of gum, oak, ash, hackberry, and an undergrowth of dogwood, vines, palmetto, haws, etc. etc. These lines of timber, reckoning that on both sides of Bayou Sale and Bayou Cypremont, is over a hundred and twenty-five miles in extent.

REMARKS.

Mr. J. C. Kathman, Chief of Bureau of Immigration, in his report, remarks: "All the upland and alluvial region, comprising three-fourths of the State of Louisiana, is covered with the finest forests in the United States; and as the State is cut up in every direction by navigable waters, the forests of pine, cypress, live oak, white oak, post oak, gum, ash, and other valuable trees, furnish employment to hundreds of mills and thousands of workmen in getting out lumber for home and foreign trade."

GENERAL ELEVATION.

In the lower or eastern part of the Parish of St. Mary, around Berwick's Bay and the lower Teche, the highest land is about ten feet above the level of the Gulf of Mexico. Near Franklin the highest bank is from twelve to thirteen feet. Near Breaux Bridge, the first bank is twenty-two feet high, the second bank twenty-seven feet.

In the Parish of Lafayette, the Cote Gelee Hills, Beau Basin and the banks of the Vermilion are forty feet above the level of the Gulf. The general average of St. Landry is about sixty feet above the same level. The Parish of Vermilion is about on a level with St. Mary.

The highest elevations on Belle Isle, Cote Blanche, Grand Cote and Petit Anse Islands, are from one hundred and sixty-two to one hundred and eighty-five feet above tide water.

PROFESSOR HILGARD'S OPINION.

Professor Hilgard, in his Preliminary Report of a Geological Survey of Western Louisiana, remarks:

"Few sections of the United States, indeed, can offer such inducements to settlers as the prairie region between the Mississippi Bottoms, the Nez Pique and Mermentau. Healthier by far than the prairies of the Northwest, fanned by the sea breeze, well watered—the scarcity of wood rendered of less moment by the blandness of the climate; and the extraordinary rapidity with which natural hedges can be grown for fences, while the exuberantly fertile soil produces both sugar cane and cotton in profusion, continuing to do so in many cases after seventy years exhaustive culture—well may the Teche country be styled by its enthusiastic inhabitants, the 'Garden of Louisiana.'"

FRUITS AND VEGETABLES.

Plums, figs, quince, pears, cherries, grapes, papaws, persimmons, pecans, hickory nuts, walnuts, blackberries, dewberries, May apples, mulberries, crab apples, black and red haws, chincapins, strawberries, and some other fruits, nuts and other fruits of little importance, thrive and mature well in these parishes.

In St. Mary, and along the coast to the Mermentau, oranges are raised yearly in great abundance, and the mespilus, or Japan plum, lemons, limes, bananas and pine apples may be produced in the open air as high up as Franklin, by giving them a little extra protection in the winter.

Turnips, cabbages, beets, and all the other garden vegetables and melons grow as well in these parishes as they do north of the Ohio river. The best winter gardens contain large white head cabbages, rutabaga and flat turnips, onions, eschallots, garlic, mustard, roquette, radishes, cauliflower, beets, cress, lettuce, parsley, leeks, English pease, celery, endive, etc., etc. These thrive well in the gardens all winter, except in very cold winters, back from the coast, when a part of the list give way before the frosts.

IRISH POTATOES.

"In 1867, Irish potatoes to the value of more than $500,000 were shipped from Louisiana to Western and Northern markets; and one farmer on the coast realized more than $25,000 from less than thirty acres of cabbages shipped to the Western States." The above we copy from Kathman's report in 1868. There is no part of Louisiana where Irish potatoes and cabbages thrive better than in Attakapas and St Landry.

OPINION OF AN ILLINOIS FARMER.

One of the largest and most intelligent farmers in Central Illinois, after a careful examination of the Teche and Attakapas country, said:

"I have heretofore thought that Central Illinois was the finest farming country in the world. I own a large farm there, with improvements equal to any in the country. I cultivate about two thousand acres in small grain, besides other crops; but since I have seen the Teche and Attakapas country I do not see how any man who has seen this country can be satisfied to live in Illinois.

"I find that I could raise everything in Louisiana that can be raised in Illinois, and that I can raise a hundred things there which cannot be raised in Illinois. I find the lands easier worked in Louisiana, infinitely richer and yielding far more; and with the fairest climate on earth, and no trouble to get to market. I shall return to Illinois, sell out, and persuade my neighbors to do the same, and return to Louisiana to spend the remainder of my days."

OPINION OF AN ILLINOIS EDITOR.

The editor of the Chicago Tribune, the leading Radical paper of the West, after visiting the Teche country, said to his 50,000 subscribers:

"If, by some supreme effort of nature, Western Louisiana, with its soil, climate and production could be taken up and transported North; to the latitude of Illinois and Indiana, and be there set down in the pathway of Eastern and Western travel, it would create a commotion that would throw the discovery of gold in California in the shade at the time of the greatest excitement. The people would rush to it in countless thousands. Every man would be intent on securing a few acres of these wonderfully productive and profitable sugar planes. These Teche lands, if in Illinois, would bring from three to five hundred dollars per acre."

BRASHEAR CITY AND MORGAN STEAMSHIPS.

The depot and wharves at Brashear have been improved by Charles Morgan, at an expense of more than a hundred thousand dollars. Mr. Morgan has six or eight splendid steamships plying regularly between this place and the ports of Texas. The largest of these ships cost three hundred thousand dollars.

RAILROADS.

The Morgan Louisiana and Texas Railroad, is in fine running order and well equipped between New Orleans and Brashear City, a distance of eighty miles. The old Opelousas Railroad, a continuation of the above railroad, is graded between Berwick's Bay and Opelousas, a distance of eighty-five miles, or one hundred and sixty-five miles from New Orleans. Mr. Charles Morgan owns the old N. O., O. & G. W. Railroad in this State and its franchise from New Orleans to Pine Prairie, and above, and from New Iberia to Orange on the Sabine.

PARISH OF ST. LANDRY.

ITS AREA AND PHYSICAL CHARACTER.

The parish of St. Landry contains about 1,350,000 acres, about equally divided between woodland and prairie. About three-quarters of the land is suitable for planting and grazing purposes. It is well watered by numerous bayous, running streams, and branches, nearly all clothed with a generous growth of timber, in many places a mile wide. The land is generally undulating, and drains well. Between the timbered streams fine natural meadows spread out, clothed over nine months in the year with grass that sustains immense herds of cattle and horses.

THE SOIL—FACE OF THE COUNTRY.

In the upper part of the parish, nearly all the streams, fed by springs, take their rise. Here the country is somewhat hilly and is covered with a dense forest of pine, oak, ash, walnut, hickory, and other valuable forest trees. Here also are found some fine mineral springs, which are much resorted to by invalids, and which possess great curative qualities. Here are considerable deposits of limestone, from which, for home consumption, is made very excellent lime; and a fine quarry of marble, which is susceptible of a beautiful polish, and is valuable for making into mantle pieces, monuments, etc.

The soil in the middle and lower portion of this parish is excellent, resting on a subsoil of fine brown or grayish clay, which, when plowed up, exposed to the weather, and mixed with surface soil is as rich as the upper stratum. That subject to overflow, being rich alluvial, is inexhaustible, and adapted to all the products of this latitude. The soil of the prairies is generally mellow, and easy of cultivation. Grass covers all portions of the parish, except the cultivated fields, or surfaces covered by forests, or by water. More than half a million acres of grass in St. Landry is not under fence. The greater portion of the wealth of St. Landry has been obtained from cattle and horses on the prairies, raised without hay or shelter. On these prairies a hundred thousand tons of hay might be made yearly for the New Orleans and other markets.

DARBY'S DESCRIPTION—PRAIRIES AND HERDS.

In Darby's Geographical Description of the State of Louisiana, published in the year 1817, when the Sabine was the western boundary of the parish of St. Landry, the following description of the Opelousas Prairie is found :

"This vast expanse of natural meadow extends seventy-five miles southwest and northeast, and is twenty-five miles wide, containing more than 1,200,000 acres, exclusive of the numerous points of woods that fringe its margin on all sides. This prairie begins thirteen miles northwest of Opelousas and gradually opening to the southward, sends out various branches between the bayous.

* * * * * * *

Of the herds as then seen on the prairies, the same author remarks :

"Here you behold those vast herds of cattle which afford subsistence to the natives, and the inhabitants of New Orleans. It is certainly one of the most agreeable views in nature to behold from a point of elevation, thousands of cattle and horses of all sizes, scattered over the intermediate mead in wild confusion. The mind feels a glow of corresponding innocent enjoyment with those useful and inoffensive animals grazing in a sea of plenty. If the active horsemen that guard us would keep their distance, fancy would transport them backwards into the pastoral ages. Allowing an animal to be produced for every five acres, more than two hundred and twenty thousand can be yearly reared and transported from this prairie alone, which, at an average of ten dollars per head would amount to two million four hundred thousand dollars."

At the time the above article was written, the year 1817, Mr. Darby estimated the herds of the three greatest stock owners of the country, Mr. Wikoff, in the Calcasieu Prairie; Mr. Fontenot, in Prairie Mamou; and Mr. Andrus, in Opelousas, at twenty thousand head.

Long after Mr. Darby's day, the Dupre and Wikoff families, and the widow Guidry owned over 30,000 head of horned cattle, and more than 6000 horses. The stock raisers of the parish owned over 100,000 head of cattle and 30,000 horses.

THE TIMBERED BOTTOMS.

The timbered bottoms are rich, and are excellent for sugar, rice, cotton, corn, sweet or Irish potatoes, pease, tobacco, melons, pumpkins, hay, gardens, fruits, etc. No richer lands can be found anywhere. They are heavily timbered with the best of sugar wood, and the swamps contain an inexhaustible supply of the best of timber for building purposes, and for hogsheads and barrels for the sugar planters.

BAYOUS, RIVERS AND STREAMS.

The Atchafalaya on the east connects this parish by steamboat navigation with the Mississippi river and New Orleans. The Bayou Courtableau, formed by the junction of the Crocodile and Bœuf, affords good navigation to Washington the entire year, with slight and occasional interruption during the summer. The route is down the Courtableau to the Atchafalaya, thence up the latter to the Mississippi river, and thence to the city of New Orleans.

The Bayou Bœuf is the channel of transportation for the planter, by means of barges to Washington; and the Crocodile affords means of transportation to the lumber men. The Plaquemine, Brule, the Mallet, the Cane, and the Nez Pique are fine streams, but not navigable. The Mermentau, formed by the Nez Pique, Cane, and Plaquemine Brule, is a fine navigable stream; vessels ascend it some 70 miles for lumber, which is taken to Texas, Havana, and the Mexican ports. Upon these streams are found large bodies of timber, suitable for all the purposes of building and fencing, and they afford an unfailing supply of water for stock.

The parish has two hundred and thirty miles of navigable waters.

OVERFLOWS.

Portions of St. Landry on the Atchafalaya and some of the bayous are subject to overflow when Grand Levee gives way ; but most of the lands have never been inundated since the parish has been inhabited by white men, and never can be.

And even the overflowed lands may be converted into rice plantations to some extent, or reclaimed when the levees of the Mississippi and Atchafalaya are made secure. Most of these lands subject to overflow are the richest in the world, and contain a heavy growth of cypress.

POPULATION AND FARMS.

In 1850 there were 775 farms in St. Landry ; 2427 dwellings ; white population, 10,189 ; free colored, 1143 ; slaves, 10,781 ; total, 22,253.

CROPS, FRUITS AND GARDENS.

The crops, fruits and gardens of St. Landry, and of the other five parishes described in this circular, excepting cotton, are less troubled by insects and vermin, and less liable to disease, than they are in higher latitudes in other parts of the United States.

The surface cultivated in St. Landry yearly, amounts to about a hundred thousand acres. About one-third of this is planted in cotton. Not a tenth part of the tillable land is under cultivation. With a working population like that of the Western States, and the same kind of cultivation, that parish might send to market yearly a hundred thousand bales of cotton, fifty thousand hogsheads of sugar, seventy-five thousand barrels of molasses and rice, tobacco, broom corn, hay, beeves, horses, milch cows, sheep, hogs, hides, poultry, eggs, rosin, turpentine, and other valuable products to the amount of from ten to fifteen millions of dollars. So varied and valuable resources, in a climate so salubrious, can hardly be found anywhere else on the face of the earth.

THE SUGAR CROP.

Small crops of sugar cane on small farms are well adapted to white labor. The cane may be planted in the fall, winter or spring, and laid by before the 1st of July ; and no labor is then needed in the crop till the first of November, when the ripe cane is ready for the mill. Sugar cane is not subject to diseases, and the ravages of bugs and insects, like most other crops. Small sugar farms, where from twenty to one hundred hogsheads of sugar are made by white labor, are very profitable. They are a complete success.

PROFITS OF SUGAR CANE AND COTTON CULTURE IN ST. LANDRY.

The following statements we copy from a pamphlet published in Opelousas, in 1869:

MOUNT HOPE, PARISH OF ST. LANDRY, January 1, 1869.

Messrs. Lewis & Mullett:

Gentlemen—I employed the past year twenty-two hands, to-wit: 15 men, 2 boys, and 5 women. Had in cane 90 acres, in corn, 170 acres; and in cotton, 100 acres, besides several acres in potatoes and gardens.

RESULT OF THE YEAR'S WORK.

Ground 58 acres of cane in 18 days, making 108 hogsheads of 1250 pounds each, which sold at 10 cents	$13,000
I made 200 barrels molasses, equal to 8000 gallons, at 70 cents	5,600
Also 7700 barrels of corn	2,100
Also 86 bales of cotton, equal to 38,000 pounds, 22 cents	8,514
Gross receipts	$29,214

EXPENSES.

My total expenses for provisions, repairs, hire of hands, sugar maker, hogsheads and barrels, were $10,000.

Which deducted from the gross income, leaves $18,719 as my year's income. I am, very respectfully, your obedient servant,

ELBERT GANTT.

P. S.—I employed 7 extra hands for 90 days.

BAYOU BŒUF, PARISH ST. LANDRY, January 1, 1869.

Messrs. Lewis & Mullett:

Gentlemen—I herewith transmit, in accordance with your request, a statement of the sugar crops produced during the year 1868, on the Barbreck and St. Peters plantations:

Sugar produced, 460 hogsheads, averaging 1250 pounds each, which, at $100 per hogshead, yields $46,000
Twenty-seven thousand six hundred gallons molasses, at 50 per gallon 13,800

Gross proceeds, ... $59,800
Expenses 19,000

Profits $40,800

Besides the above, we produced fifteen thousand bushels of corn, which is at least three or four thousand more than we require for the use of the plantations. Very respectfully, etc.,

H. M. Payne.

Labyche Plantation, January 5, 1869.

Messrs. Lewis & Mullett:

Gentlemen—Mr. Gantt handed me your letter last evening ;the following is a correct statement of the crop asked for :

CROP OF LABYCHE PLANTATION.

Ninety-two acres of plant cane and 82 acres of corn, cultivated by twelve men, produced 193 hogsheads of sugar, 341 barrels of molasses, and 1600 bushels of corn.

We ground 72 acres of cane, and sold 3 acres for $1,000.

The 72 acres produced 193 hogsheads of sugar weighing $28,090 pounds ; 341 barrels of molasses, 14,322 gallons ; averaging per acre 3029 pounds of sugar, and 200 gallons of molasses.

The net proceeds from the above, including 3 acres of cane . . $32,221 18

EXPENSES.

For cultivating crop $1,235 00
For taking off the same 1,551 31
For pork, in lieu of wages 662 00
Sugar hogsheads, molasses barrels, repairs to sugar house, and other small expenses .... 3,500 00 — 6,948 31

Net proceeds $25,272 87

We consumed in making the sugar, 660 cords of gum wood ; in taking off the crop we hired 30 extra hands ; their wages is included in the $1,551 31. The number of days cutting cane, 42. The number of days grinding cane, 21. Total capital invested, $40,000 00.

Your obedient servant,

GEORGE W. MORGAN.

YIELD OF COTTON, SUGAR AND OTHER CROPS.

In St. Landry, 1300 pounds of seed cotton to the acre, or about 400 pounds of lint, is a fair yield. In the true cotton zone, which is above the latitude of this parish, about 32 degrees north, 1800 pounds of seed cotton may be produced, or 600 pounds of lint.

Whilst it is admitted that the cotton plant is liable to injury from insects, still, in the main, as many full crops are made as of any other product of the soil, and the chances of success are by many thought to be as favorable in this branch of industry as in any which engages the farmer.

One hogshead of sugar and sixty gallons of molasses may be considered an ordinary yield per acre in this parish ; but we are assured by a gentleman for whose veracity we have high respect, that twenty-five hogsheads of sugar have been produced in St. Landry from six acres of ground. That is the best yield that has ever been known in any sugar parish in the State. Sixty gallons of molasses usually drain from a hogshead of sugar. Commercial manures will doubtless largely increase the average yield of sugar in all of these parishes, and the same facts hold good in regard to cotton, and other crops.

The yield of corn in St. Landry, is about 35 bushels to the acre. Potatoes, sweet and Irish, well cultivated, from 250 to 300 bushels to the acre. Pumpkins, peas, beans, pindars, broom corn, etc., etc., give heavy returns, but, owing to the fact that no one ever made a note of the yield of these crops per acre, correct statistics of them have never been recorded.

PRICE OF LANDS.

Land near the navigable streams and towns can be purchased at from $10 to $15 per acre ; at a distance from the centre of business, at $8 and $6 per acre ; in the western portion of the parish, at $1 25 and $2 50.

There is much public land subject to entry under the Homestead Act, in tracts of 80 or 160 acres, costing, (the surveying and incidental expenses included), $23 50 per each 80, and $45 for 160 acres; much of this latter quality is covered with fine timber, and near large water courses. Every class and condition will be enabled to find lands adapted to their several tastes and pursuits. The capitalist can purchase large bodies of land from private persons for the purpose of colonizing, and the man of small means, a farm adapted to his means and wants. Saw mills are scattered through the parish, enabling all to secure lumber for building purposes, at prices ranging from $20 to $35 per thousand feet. Corn will abound in the parish, and the immigrant will thus be enabled to secure bread for the year.

THE TRADE OF WASHINGTON.

About 8000 bales of cotton have been shipped from Washington and Barry's Landing, (Port Barre) up to the 22d of January, '70. There are yet about 12,000 bales to be shipped. The house of Pitne & Carriere shipped in twelve months, to September 1869, 6307 hides, 33 rolls, 3220 pounds of leather; 53 bales of wool, and 2597 bales of cotton. But the cotton crop of that year was very short.

They ship from 1200 to 1800 boxes of eggs to New Orleans yearly, ranging from twenty to eighty dozen in a box. Offut & Stagg did about the same business last year.

They ship yearly to New Orleans from St. Landry about 3000 boxes of eggs, 2000 coops of chickens, turkeys, ducks, etc. Say 150,000 dozen eggs, generally sold in New Orleans market at 40 cents a dozen, $60,000; 2000 coops of chickens, etc., 5 dozen in a coop, 10,000, at an average of $4 a dozen, $40,000; equal to $100,000 a year for the chickens and eggs of St. Landry, sold. The real value of the yearly product of those articles in St. Landry cannot be short of $200,000.

FREIGHTS, ETC.

The steamboat charges on cotton to New Orleans, $2; sugar, $4; molasses, $1 50; coops of chickens, $2 50; eggs, 2 cents a dozen; hides, 10 cents each; up freights, 75 cents for dry barrels; $1 for wet barrels; passage to and from New Orleans, $10 each way.

DISTANCES.

From Washington to Barry's Landing, or Port Barre, 12 miles; to wharf Boat, at the mouth of Red River, 117 miles; Wharf Boat to New Orleans, 210. Washington to New Orleans, 339 miles.

Time between New Orleans and Washington, 35 to 40 hours. They live high on these boats; their tables are loaded with luxuries of the New Orleans market, much like the tables on the Attakapas boats in former years.

THE CROPS OF 1869.

The cotton crop of St. Landry last year was about 20,000 bales. But little sugar was made. The crop was short, and the planters have been more inclined to cultivate cotton than sugar since the war. Not more than 1000 hogsheads of sugar, and 1500 barrels of molasses were made in the parish. Most of the cotton planters made nearly a bale of cotton to the acre.

WHITE LABOR—PROFITS.

On Col. Roger's place, three miles from Opelousas, a white man and boy made 18 bales of cotton; two white men and two boys made 35 bales; and one white man and a boy made 16 bales. All the hands on the place, 8 in number, are white, and all the domestic work, cooking, washing, milking, etc., is done by white people.

Mr. Hypolite Davide, residing near Grand Coteau, with the assistance of one negro boy, eighteen years old, made last year, thirty bales of cotton and fourteen wagon loads of corn. We have the above statement from Mr. Dunbar, postmaster of Grand Coteau. It is not understood that the two hands picked and saved the crop. They only made it, and must have had help to save it.

BELLEVUE AND WHITE LABOR.

Most all of the agricultural labor of Bellevue is done by white people. This section of the parish is south of Opelousas. There are six families from Mississippi, three from Kentucky and three from Arkansas, cultivating small places here. They are a hard working people, and will succeed. Most of them eat breakfast by candle light, on the north Alabama plan, and rush business during the day. In cotton picking they take their dinners in the field. They practice rigid economy. Even some of the white women pick cotton. A white boy, fourteen years old, picked three hundred pounds of cotton in a day, when they were paying hands one dollar a hundred and boarded. In some places in the parish, hands could not be obtained to pick all of the crop, and there was considerable loss.

A circular published in 1869, endorsed by the leading and most intelligent citizens of St. Landry, members of the Agricultural Society of that parish, states :

1. White men have tilled the soil of this parish from the early settlement of this State, and are now tilling it without experiencing any serious difficulties.

2. One white man easily cultivates 20 acres of land here under the old system, and much more by the use of labor saving machinery lately introduced, and from his individual exertion he can reap a larger return than in the Western and Middle States.

3. Out door labor to the white man is not unpleasant, and can be sustained without detriment to health. The average temperature of the winter is about 40 degrees, and that of the summer 85 degrees during the day. During the heated period, we have none of the sultriness of the more Northern States at night. This has struck every one as a remarkable feature in our climate. After the heat of the day there comes a delicious feeling of pleasure, produced by the cool sea breeze and the refreshing dew, which combine to cool the hot earth, tone and invigorate the system, and prepare it for the coming day.

4. More than one-half of the white population of this parish are engaged strictly in agricultural pursuits, and are as robust and healthy as any similar number of farmers in higher latitudes.

5. The production of one white man's labor, who is industrious, may be thus stated : 400 bushels of corn ; 200 bushels of sweet, and as many of Irish potatoes ; 10 barrels of rice ; 5 bales of cotton ; with a full supply of vegetables, hay, etc. This statement is the actual result of one man's exertions, and may be safely taken as a fair average.

We appeal to the experience of five hundred farmers in our midst for the truthfulness of every fact given, and do not fear contradiction in one single particular.

The same document, treating of the cultivation of sugar cane by white labor, says :

White men can insure to themselves a fair revenue by the cultivation of sugar cane upon a small scale, the molasses paying the current expenses.

It opens the way for co-operative associations. The man of means, by putting up the machinery, can thus secure to the small farmers near him the means of saving their crops, and thus stimulate the enterprise of the people. His investment will return to him a good interest, and at the same time promote a healthful division of labor.

The method here indicated has been tested in our section with excellent results, and encourage further efforts in the same direction.

DR. TAYLOR'S STATEMENT.

The following statements were furnished us by Dr. Taylor:

Last year, 1869, Mr. Lemorardier, assisted by his brother-in-law, made 21 bales of cotton, a large crop of corn, and performed all the labor of the farm, attending to stock, gardening, etc.

Mr. Dunbar rented the Eureka plantation of W. O. Denegre, Esq., of New Orleans. His four sons, the youngest 16, his son-in-law, Mr. Simpson, and two negro men, seven in all, made 77 bales of cotton, and 1200 bushels of corn. It is worthy of remark that none of these white men were accustomed to farm labor.

Mr. Dunbar sub-rented portions of the plantation to Messrs. Humble and four Germans.

The two Messrs. Humble, with their two sons, one 12, the other 13 years of age, one negro boy and a negro man, made 70 bales of cotton, 600 bushels of corn, and cultivated and sold 200 bushels of Irish potatoes, and cultivated three acres of cane, which was sold for $450. They hired some labor in scraping the cotton, and had some assistance in picking.

Mr. Duckworth, with one young man and two boys, made 32 bales of cotton, and an abundant crop of corn. Mr. Duckworth did but little himself, so the crop was made chiefly by the young man and boys.

The Messrs. Gilette, father and son, in addition to a corn and cotton crop, cultivated some sugar cane. They made 12 bales of cotton, and 16 barrels of syrup on a small sorghum pan. The younger Mr. Gillette is a promising young minister in the Methodist Church, and attending during the year to his clerical duties,

These examples prove that white labor is eminently successful in this parish. I don't believe there was a case of sickness among these people during the planting, cultivating, or harvesting season. All experience has demonstrated that while the white laboring man in this climate enjoys equal exemption with the negro from disease, his labor, from his superior intelligence and energy, is much more remunerative.

I have always found white laborers who work in the field in St.

Landry the year round, as healthy as field laborers in any other States, as healthy as white men who do not labor; and that this fact holds good in regard to white men who were raised in affluence and ease, and who have done their first field work since the war.

J. A. TAYLOR, M. D.

MAST AND HOGS.

St. Landay abounds in oak forests, and in mast of various kinds. The hog range is excellent, and white clover grows luxuriantly, equal to the native grasses. This is fine for grass eating hogs. And in no portion of the United States are hogs more healthy, or more prolific than in St. Landry.

OAK BARK AND TANNERIES.

In a country abounding, as St. Landry does, in oak bark, sumac, and hides, and where tan pits may be kept open during the entire winter, what is there in the way of success in tanneries as the country becomes older, and labor becomes more diversified?

FEEDING STOCK—PROFITS.

A stock raiser of St. Landry writes to the Agricultural Society of that parish as follows:

"I am a stock raiser. It is necessary to feed gentle cattle from the 15th of January to the 15th of March. In mild winters very little food is necessary. Wild stock never are fed on hay, and have no shelter. Wild stock yield twenty-five per cent. income on the investment. Gentle stock may be made to yield over forty per cent.

WELLS.

Well water in St. Landry is cool, and wholesome. Average depth of wells, 25 feet; sometimes fifty feet. There are a few cisterns in the parish, but most people use well water. The wells are usually curbed with cypress.

CROPS ADAPTED TO ST. LANDRY.

Cotton, corn, sugar cane, broom corn, ramie, flax, hemp, sweet and Irish potatoes, cow pease, indigo, sorghum, pindars, castor oil beans, oats, barley, rye, pumpkins, cabbage, turnips, and garden vegetables of all kinds.

FRUITS.

Peaches, apples, pears, plums, figs, grapes, quince, blackberries, strawberries, dewberries, may apple, persimmon, may haw, and papaw. Oranges may be successful to some extent in the southern part of the parish. But little attention is paid to fruit culture.

DOMESTIC ANIMALS.

Cattle, horses, mules, jacks and jennies, sheep, hogs, and goats.

DOMESTIC FOWLS.

Turkies, chickens, geese, ducks, guinea fowls and pea fowls.

FISHES.

Perch, trout, buffalo, pike, white cat, sac-a-lait, or white perch, mudfish, or choupique, soft shell turtle, and in some of the bayous, the finest trout in great abundance.

WILD ANIMALS.

Deer, bear, panther, opossum, raccoon, rabbit, wild cat, fox, otter, and squirrel.

WILD GAME.

Wild turkeys, duck, geese, brent, papabots, snipe, partridges, prairie chickens, reed birds, grosbec, wood cock, rice birds, and robbins.

DOMESTIC ANIMALS—HOME PRICES.

Creole horses, $40 to $60 ; formerly $20 to $30 ; beeves $12 to $20; milch cows $20 to $30 ; sheep $1 50 to $2 ; beef in the market, 5 to 8 cents a pound.

DOMESTIC FOWLS—HOME PRICES.

Gobblers, $1 50 ; hen turkeys, 75 cents to $1 ; geese, $1 a pair ; chickens, $2 50 to $3 a dozen ; ducks, 25 cents ; wild ducks, 20 to 25 cents ; eggs 8 to 10 cents in summer.

FARM CROPS AND VARIOUS ARTICLES—HOME PRICES.

Corn, 75 cents down to 25 cents a bushel ; butter, 40 cents in winter, 25 cents in summer ; sweet potatoes, 30 to 50 cents a bushel ; Irish potatoes, $1 a bushel ; cypress lumber, $25 to $40 a thousand feet ; cypress shingles, $5 to $8 a thousand ; bricks, $10 to 12—can be made at $7 ; fencing, pieux and posts, $80 to $150 a thousand. (It takes 5 pieux and one post to a pannel of fencing 8 feet in length, and nearly 4000 to a mile).

CATTLE, HORSES, ETC.—NUMBER.

The number of cattle in St. Landry in the year 1869, is estimated at 60,000 head; horses, 15,000; mules, 3500; sheep, 3000; number of hogs large, but not known.

CHILDREN AND YOUNG PEOPLE.

The number of persons, white and black, in this parish, between the ages of 6 and 21 years, amounts to 8380, classified as follows: White males, 2,490; females, 2,344; total, 4832; colored males, 1823, females, 1725, total, 3548.

LONGEVITY IN ST. LANDRY.

By the assistance of intelligent citizens of St. Landry, we have a list of names of white persons in Opelousas and the parish, above the age of sixty-five years. We find that there are twenty-two in Opelousas and fifty-six in other parts of the parish, making seventy-eight persons in the parish of St. Landry above the age of sixty-five, the oldest having arrived at the remarkable age of one hundred and eighteen years. In Opelousas there are forty-one white persons between the ages of fifty and sixty-five years. And doubtless there are many others who belong in the list, but we have not been able to obtain their names. There are many old colored people in this parish, some of them very old.

THE WIDOW DUFRENE.

The widow V. Dufrene lives about fifteen miles from Opelousas. She has arrived at the age of one hundred and eighteen years. She weighs less than a hundred pounds, is tall, straight, has a very good eye-sight, and walks briskly for one of her advanced age; and her mind has a fair amount of vigor.

STILL OLDER.

Joseph Cheasson, alias Joannes, died several years ago in this parish at the advanced age of nearly a hundred and thirty years. When he was one hundred and fifteen years old he moved to Texas, and after living in that State several years he returned to St. Landry.

Mr. Thomas died in this parish several years since aged over a hundred years.

Joseph Young died in this parish thirty years ago, aged about one hundred and fifteen years. He married at the age of ninety years, and his wife had a son whom he lived to see married. His widow still lives in St. Landry.

Mrs. Blaize died in this parish a few years since, aged more than one hundred years.

Mrs. Daigle, aunt of Mr. Choteau, who has a lease of the Avery Salt mines, died in Opelousas. She was nearly one hundred years old.

Jesse Andrus, aged ninety years, and Major John Clac, aged ninety, lately died in this parish.

A NUMEROUS PROGENY.

A respectable physician, of Washington, informed us that Madame Guillory, an old lady of St. Landry, before her death, could count up over eight huudred lineal descendants, all blood relations of hers.

A HEALTHY FAMILY.

Mr. Joseph Langley, 95 years old, lived in St. Landry 42 years; his twelve children, all of them are living. He had no physician in his family for twenty-five years. When over 90 years old, in aiding a deputy sheriff who wished to cross a swimming creek, throwing his feet up by the pummel of the saddle he crossed without wetting them, while the young man who followed him wet both feet and half of his legs.

PARISH OF IBERIA.

GENERAL DESCRIPTION.

Iberia parish extends from Belle river, east of Grand Lake, to a line running from the west end of Lake Peigneur to the mouth of Petite Anse bayou. It is bounded on the north by St. Martin, and on the South by St. Mary, east by Assumption, and West by Vermillion and Lafayette.

Its length is about forty-five miles, its widest part is about twenty miles. Much of the eastern portion is water and cypress swamp. The tillable land along the west side of the Morgan Railroad and the Teche, from the parish line below Jeannerette to New Iberia, called the "Prairie au Large," has a width of about six miles, and it is a little wider above between the railroad and Lake Peigneur; the land from the line where the railroad enters the parish below Jeannerette to the line where it leaves it west of Lake Tasse, is about twenty miles in extent. All the Land is tillable between Lake Peigneur and Lake Tasse, and in the great bend of the Teche northeast of New Iberia. And there is a sheet of tillable and fine grazing land south of Lake Peigneur.

The Teche is lined with plantations nearly the entire distance from its entrance into the parish of Iberia east of Lake Tasse, to the line where it leaves the parish below Jeannerette. The portion of the parish that borders on Grand Lake is a dense cypress swamp, and bordering on this swamp there is a growth of gum, ash, oak and other timber. The tillable land opposite and above Jeannerette is two or three miles in width. Around the great bend of the bayou above, called Fausse Pointe, the tillable land has a much greater width. The lands in all parts of this parish are rich. On the west side of the bayou there is a scarcity of woodland; on the east side there is an abundance of fine cypress and wood for sugar making.

THE TECHE AND ITS SCENERY.

From the point where the Teche enters the parish of Iberia, about five miles below St. Martinville, by the windings of the bayou, to New Iberia, the distance is about twenty-five miles. This portion of the bayou is extremely beautiful. Its banks are generally about eighteen feet above the water, and they descend gently to the edge of the water at an angle of less than thirty degrees. The bayou, around this bend, in the low water season, is about ninety feet wide, and has a depth on its most shallow bars, of about three and a half feet. Forest trees and water willows line both banks most of the distance, the branches in many places hanging over the water, and brushing both wheelhouses of steamers as they pass up and down between New Iberia and St. Martinville.

The houses of the planters and small farmers are generally situated not a hundred yards from the edge of the bayou. Most of the houses are plain, but comfortable, and the improvements generally are plain; but the proprietors are quite independent.

There are many large, fine live oaks, pecans and other noble forest trees growing on both banks of the bayou, that add greatly to the beauty of this section of the parish.

Below New Iberia the Teche is broader and deeper than above, the plantations are larger, the houses and improvements finer, and there are fewer trees growing on its banks. Here we find palatial residences, grand sugar houses with chimneys towering skyward, plantation villages called "the quarters," orange groves, groves of the mespilus, flower gardens and beautiful shrubbery, floating bridges, and the general parahernalia of wealth and lordly possessions. Yet, on this part of the bayou there are many plain or unsightly buildings, and some plantations that have not yet got rid of the footprints of the late war.

THE PRAIRIE AU LARGE.

This is the sheet of land south and west of New Iberia; and a more beautiful prairie country is seldom or never seen. It has natural drains, which, by being opened a little, would relieve the whole country from surface water after rains. Leading natural ditches penetrate all parts of the prairie, and into these the ravines may be opened at small expense. This sheet of fertile prairie must, at no distant day, be put in a high state of cultivation by small farmers. Though there are many thrifty little fields now under fence, we doubt if a tenth part of this prairie is cultivated. Small tracts of from forty to two hundred acres can be bought for ten dollars an acre and even less. Large planters cannot come into this prairie and put up new and expensive machinery with any prospects of success. A small farmer can start with cheap improvements, make ten, twenty, or fifty hogsheads of sugar yearly, with a certainty of success. His coal may be hauled from the banks of the Teche at his leisure at any season of the year in dry weather. A ton of coal, at a cost of $6 50 on the bayou, will boil a hogshead of sugar.

In addition to the sugar crop the small farmer could raise milch cows for sale and make butter and cheese for the New Orleans market. And poultry, eggs, garden vegetables, fresh pork, broom corn, pindars, corn, potatoes, hay, melons, fruits, and other productions may all be sold for ready money or goods at New Iberia, or in New Orleans, which is but about 120 miles distant by railroad, 80 miles of which is in running order, the balance graded.

A TRIP.

In November, 1869, we took a trip in the parish of Iberia, passing round Fausse Pointe, and little Fausse Pointe. We copy from our account of the trip published in the *Banner* of December 1, 1869:

AROUND NEW IBERIA.

The more we circulate over this country, of which New Iberia is the trading centre, the more its real value and innate worth are elevated in our estimation, It is a lovely and wonderful country. Its bayous lakes, prairies and woodlands are all beautiful. Its soil is rich, deep and inexhaustible. Sea breezes roll over it, and give health and long life to its inhabitants. Its climate is a medium between the tropical and the north temperate, combining most of the advantages of both, and the evils of neither. Steamers from New Orleans, and vessels from the ocean penetrate to its very centre. And the cars of the Southern Pacific Railroad, connecting New Orleans and the Pacific Ocean, will in a few years pass over it.

SUGAR HOUSE OF DUCLEON BONIN.

In company with Dr. Shaw, we called at the sugar house of Ducleon Bonin, across the bayou, twelve miles from New Iberia. The sugar house was made of pieux and rough plank, dirt floor, everything rough and cheap. The sugar house and mill house cost $650; the mill, second hand, 32-inch cylynder, cost $500; the kettles, capacity for two hogsheads in twenty-four hours, cost $150, second hand; the whole cost of all, $1250. They will make 45 hogsheads of sugar and 60 barrels of molasses, worth over $5000. They have made two hogsheads to the acre from stubble cane. They make over 600 barrels of corn. They put up twelye acres of seed cane.

A CHEAP SUGAR HOUSE.

The three Bonin brothers, Martial, Valcourt and Ovignac, were raised in Fausse Pointe, served through the war in the Confederate army, lost all their negroes and nearly all their other property, but one hundred and seventy acres of land where they now live. Last year they went into the swamp, cut the timber and floated it out with their own hands, made their pieux, and ten or a dozen neighboring Creoles joining them they put up their pieux sugar house in one day. The sugar house entire cost no money except the expense of a keg of nails; the house has a dirt floor to it; the molasses drains so as to catch it in an old sugar kettle, and from this it is barreled for market.

CREOLE INDUSTRY AND PERSEVERANCE.

The three Bonin brothers will this year make thirty hogsheads of sugar. They are now making two hogsheads to the acre, and will soon make two and a half. And they will have two hundred barrels of corn to sell. They cultivated their crop with Creole horses of their own. This is what white Creole labor can do in the cultivation of sugar.

PIERRE BROUSSARD.

Our interview with Mr. Broussard, in company with Dr. Shaw, was interesting. Mr. Broussard was born at Grande Pointe, parish of St. Martin, in the summer of 1777. He descended from the Acadians who were driven from Canada by the outrageous oppression of the British government. The Chittemaches Indians were a flourishing tribe when he was young. The chief was Soulier Rouge. Mr. Broussard knew him well. The warriors then had bows and arrows with flint heads. Much of the food of these Indians was muscles, which they found in the lakes, bays and bayous.

Mr. Broussard was well acquainted with John Hays (Jack Hays, as he called him), when he first came to Petite Anse Island, in 1790, seventy-nine years ago.

ROUND THE POINTE.

From Mr. Broussard's we proceeded down through the Little Fausse Pointe Settlement to Wyche's Crossing. This is a fine settlement, principally of small farmers. They raise cotton and sugar. The cotton looks fine, and they are all doing well. From the crossing at New Iberia to Wyche's crossing, around Fausse Pointe and Little Fausse Pointe, by the route we came, the distance is twenty-four miles. From Wyche's crossing, in an air line, to the New Iberia crossing, the distance is but two and a half miles. The land in the bend of the bayou is principally owned by Dr. Duperier. And the beauty of this country, to the right and to the left, in front and behind, as we traveled under a clear sky and over a pretty good road, we cannot correctly describe. The fine prairie, the rich and generous soil, the cotton and sugar fields, the forests near at hand, and the numerous points of timber in the distance, are all enchanting to any one who can appreciate one of the finest countries that the Creator in His goodness ever formed and fashioned out of chaos.

GRAND COTE ISLAND.

In August, 1866. we visited Week's or Grand Cote Island, and gave an account of our visit in the Planters' Banner, from which we make the following extract:

The next morning after breakfast Mr. Weeks rode over the Island with us. It is somewhat larger than Cote Blanche Island, being two miles in diameter; and nearly round. As much as we have heard of the beauty of this Island, it is far more beautiful than we supposed it to be.

On one of the bluffs of the Island we had a fine view of outspread fields of cotton and corn, just in their glory—ravines, vallies, hillsides. and level plains, timber and open lands, canebrakes and pastures. In one direction is a bold elevation covered with a heavy growth of timber, and hillsides steep as the mountains of Scotland; in another direction, way down below us, between steep elevations, a pure, fresh-water lake is spread out, with water-lilies upon its surface, the branches of beautiful forest trees extending far out over the surface. It needed but a few white swans to complete the picture, and make it perfectly enchanting.

Leaving this delightful place, we piunged into the thicket between the rugged hillside and the lake, traveling in a road over which the wild cane hung so low as to make us bow to the pummels of our saddles to save our hats twenty or more times in half that number of minutes. Emerging into a valley between hills which formed a great basin, we came to a fine field of short staple cotton, like that we had seen before, clean, rank and well bolled. Coming out of this valley we mounted to higher regions, and proceeded towards the dwelline house and gardens. On our way, to our right, we noticed another fresh-water lake, having an area of about an acre.

The dweliing house is on a handsome bluff of regular shape, about one hundred and fifty feet above the level of the Gulf. Beautiful shade trees and the sea breeze keep the yard and the house cool, even in the hottest summer days. The yard all round is well set in Bermuda grass. In front the sea marsh extends out a hundred yards, and beyond this the waters of the Gulf spread out under a blazing sun. To the right is a bayou twenty feet deep, with five feet of water on the bar at its mouth. Any of our bayou steamers can run up to the landing, a few hundred yards from the dwelling house. Redfish, and many other fine fishes are found in abundance in this bayou.

The waters in front abound with fine fishes, which in the proper seasou, are taken in great abundance in the seine. There are oyster reefs not far off.

In the garden we feund a splendid arber of Scuppernong grape vines, about thirty feet squäre, the roof about nine feet high, the vines flowing down to the ground on all sides, making a complete room with vegetable walls and ceiling. The vines on all sides and above, hung full of grapes not yet ripe. These vines produce a bountiful crop of grapes every year. There is no doubt that all this chain of islands on the coast, Petite Anse, Grande Coto, Cote Blanche, and Belle Isle, are admirably adapted to grape culture, and will at some future day become as celebrated for their wines as the islands of any portion of Europe.

Fruits of all kinds appear to do well on all of these islands.

This Island contains a surface of about two thousand acres, six hundred acres of which are in timber, the balance is in pasture, or under cultivation.

Any one may visit either of the three Islands, Grande Cote, Cote Blanche. or Petite Anse, in a buggy, and when not too wet, the road will be found pretty good, and always entirely safe.

At various localities, all over the Island, fine, thrifty forest trees may be seen, which add much to the beauty of the scenery. The Island, viewed from its highest pinnacle, is picturesque and beautiful beyond anything we have ever seen in the State. Its gentle undulations, its peaks, hills, valleys, ponds, its towering magnolias and noble oaks, its ash and cypress, its fields of blooming cotton and waving cane, all inspire the most pleasant emotions in the breast of any beholder who loves to look on nature when she puts on her finest robes and appears in her most bewitching mood.

The Weeks plantation, now under a high state of cultivation, on this Island, has on it all the buildings and improvements common to the largest and most successful sugar estates in Attakapas—a large brick sugar-house, slate roof, and a powerful eugine and sugar mill, capacity for taking off and saving six or eight hundred hogsheads of sugar yearly; the plantation is in fine condition, the soil is of unsurpassed fertility and the estate has always been one of the most productive and suacessful in this section of Louisiana,

PETITE ANSE ISLAND.

The same month we visited Petite Anse Island, and gave the following account of it in the BANNER:

This is called "Salt Island," "Avery's Island," "Petite Anse Island," according to the caprice of the speaker or writer. It contains about 2200 arpents of upland and 1200 arpents of timber—cypress, gum, magnolia, oak, etc. It is about two miles in diameter, nearly round, and is made of hills, valleys, ravines, ponds, woodlands, open fields and pastures, the whole surrounded on all sides by sea marsh, which, in the distance, has the appearance of dry, level prairie.

A PISGAH VIEW.

From the sugar house we climbed a steep hill to the highest elevation on the Island. On horseback, our eyes were on a level of about 100 feet above the tide water. Here we saw the Vermillion timber, and the line of timber on the Teche, apparently extending from a point below the Jeannerette neighborhood, to the neighborhood above St, Martinville. We could see Vermilion Bay, Grand Cote Island, Cypremort woods, a sail and the smoke of a Morgan steamer on the Gulf. The lighthouse could bh seen with a glass, 28 miles distant. The vast sheet of sea marsh which surrounds the Island, the fine expanse of prairie back of New Iberia and the Teche, and on towards Vermillion were all before us. Our eye swept a circumference of over a hundred and fifty miles, and a diameter of about fifty.

Could all the vast sheet of prairie and woodland, marsh and water, hill and valley, be placed on canvas, by a skillful artist, it would give to the admirer of Nature's charms a world of beauty in a nutshell.

THE SALT MINE.

The Salt Company has lately been prospecting with reference to the extent of the salt mine of this Island. By boriug they have proved that the bed is half a mile square, and it may extend a mile or more. They have gone over 65 feet into the solid salt, and find no signs of the bottom of the stratum. It doubtless extends hundreds of feet below the surface of the Gulf. The surface of the salt is about on a level with tide water, and forms a level plane, with a few slight inequalities on the surface. The earth covers the salt from eleven to twenty-five or thirty feet; and if the mine extends under the hills, they may find it covered in places a hundred feet or more in depth.

BONES, BASKETS, ETC.

Near the surface of the salt, or not many feet above it, the miners have found the bones of mastodons, and Indian baskets, made of wild cane, like those of the present day; the shape of the baskets was that of bags, and nearly the size of a corn sack; they were in as perfect a state of preservation as though they were made yesterday. These baskets and bones have been laying side by side, beyond doubt, more than five hundred years; perhaps they are as old as the mummies of Egypt. Part of a mortar was found near the baskets, and pottery also. An old Indian pottery kiln was found not far off, with some whole pots in it. Numerous flint arrow-heads are found in different parts of the Island.

AN OLD RESIDENT.

We visited Mr. John Hays while on this Island. He had lived there for 77 years. He was born in Pennsylvania. He said the Island was a dense forest when he first came here, in the year 1790. There were no Indians on the Island then, and no signs that they had been there for ages. He could never get an Indian to go on the Island. He wished them to go there to kill bears, panthers and wild cats, but they said a great calamity once happened to the Indians there, and they had never dared visit it since,

LOUISIANA ROCK SALT.

The N. O. Times gives the following notice of the samples of Louisiana rock salt from Petite Anse Island, seen at the late State Fair in New Orleans.

The committee showed deep interest in the Louisiana rock salt from near Vermillion Bay, which has been fully proved to be superior to Liverpool, Turks Island or any other salt found in the Southern market. It is sold cheaply at New Orleans. The delegates refer specially to it because of the importance of saving meat in Texas, "which is erroneously believed to be more difficult than in other States, whereas the difficulty in saving, and the frequent loss of meat, are clearly owing to the inferior, deliquescent salt imported into our States.

The committee were doubtless aware of the fact, and will report, or have reported accordingly, that by Professor Hilgard's analysis, and that of Professor Jones, the purity of this salt stands at 99.617 per cent., having less than one per cent. of foreign matter in it.

VISIT TO THE SALT MINE.

On the 16th of May, 1870, we visited Petite Anse Island, and made the following note of the fact in the BANNER of the 18th:

THE SALT MINE RAILROAD.

From the landing on Petite Anse Bayou to the salt pit the distance

5

is a mile and a quarter. A nice, new railroad is nearly completed, connecting the landing with the mine. Here steamers will take in cargoes of salt for New Orleans, and perhaps for the Mississippi and Ohio. The distance to the open Bay through this bayou is about nine miles. There is from six to seven feet of water in this line of bays to the mouth of the Atchafalaya, and from the South West Pass of Vermillion Bay to the Atchafalaya.

Mr. Choteau has barges on their way to the mines. intended to take, each of them, about 250 tons of salt to Brashear, or up the rivers, on three to four feet of water. Some of these barges will soon be at the salt landing on Petite Anse Bayou.

THE SALT PIT.

Last Monday evening we, with five others, went down into the salt pit, in the very bowels of Petite Anse Island. We stood on a platform about six feet square, which was suspended by a two-inch manilla rope. This rope and platform are worked by a steam engine. Standing on the platform, the signal given, we dropped down the depth of eighty feet, through fifty-eight feet of solid rock salt, in half a minute. Here we were about sixty feet below the level of the Gulf of Mexico, in the heart of the salt mine.

From the platform, the chamber extends two hundred feet west, and a hundred and fifty feet east, making a continuous chamber three hundred and fifty feet long, twenty-five to thirty feet wide, and seven feet high. The miners were at work. They have worked twenty-two hands and can soon make room for fifty hands. Three men are getting out about six tons of salt daily—two tons each.

LAKE PEIGNEUR.

Lake Peigneur is sometimes called Lake Simonette. It is situated about nine miles west of New Iberia. The length of this lake is about three miles, its width one mile, and its greatest depth thirty-two feet. It is fed by numerous springs that break out around its margin, It is the most beautiful lake in Attakapas, and in the South. It swarms with fishes such as abound in the lakes and bayous of this country, and they can be caught in unlimited abundance at all seasons of the year. The supply can never be exhausted.

This lake has an elevation nine feet above the surface of the Gulf of Mexico, and is situated ten miles north of Vermillion Bay, and about six miles northwest from the salt mines of Petite Anse Island. The country around this lake is very picturesque.

ORANGE ISLAND.

Formerly Miller's Island, bounds Lake Peigneur on the south. The Island lies in the curve of the lake which has the shape of a new moon. The highest part of the Island is seventy-five feet above the level of the Lake, and eighty-four feet above the level of the Gulf. It has hills, valleys, level and inclined planes, and from its bluff banks in places, the branches of the trees hang out over the waters of the lake.

Orange Island is in a line with Petite Anse, Grande Cote, and Cote Blanche Islands, and each is separated from the neighboring Islands by a distance of about six miles. Orange Island rises above the lake and the surrounding prairie, as the other islands rise above and overlook the surrounding sea marsh.

But a short distance off flows the Petite Anse Bayou, draining the neighboring country, and emptying into the Gulf ten miles below the Island. A constant sea breeze renders this spot healthy and pleasant as a residence.

There were formerly six thousand orange trees on this Island, bearing an immense crop of oranges yearly. Most of these are still in fine condition. Many of these trees have large, fine bodies, a foot in diameter, and they are more than thirty years old.

There are over 2000 young and bearing pecan trees, a large number of the better kinds of cherry, and some fig, peach, quince, mespilus, lemon and plum trees—several avenues of live oak, and other growth, and a grove of stately magnolias. Seen from the summit of the bluff, Lake Peigneur spreads out almost beneath the feet of the observer, while the gleam of its silvery surface closes the vista of the principal avenues leading from the house.

The owner of this property is Mr. Joseph Jefferson, a most genial gentleman and keen sportsman, whose renown as one of the brightest luminaries of the stage is world-wide. Mr. Jefferson contemplates spending from his ample fortune a large sum of money in the improvement of his place. This he can well afford, as his talents bring him an income of over $100.000 a year, and his property, it is said, amounts to a million.

With proper taste and judgment few places are susceptible of higher improvement in point of profit or beauty. The value of the orchard may be only partially understood by its yield of 500,000 oranges the past season, after a neglect of ten years. These readily sell at a dollar a hundred on the ground—the fruit gathered by the buyer. The beauty of the bold banks and knobs presented to the lake shore, on the north, or the natural lawns gently sloping to the south for miles, form scenic beauties attractive to poet or painter. A visit to this beautiful spot is necessary to its appreciation.

LAKE TASSE.

Lake Tasse is sometimes called Spanish Lake. Its lower limits come within two miles of New Iberia. Its length is nearly five miles, its shape is nearly round. One-third of the surface is water, and two-third grass, usually called floating prairie. It has generally been supposed that nearly the whole of this grass surface consists of matted roots a foot or two in thickness with water underneath. This proves to be an error. Some of the grass next to the water surface is afloat, and fishermen run on the surface with skiffs, work holes through with their paddles, and catch immense quantities of fine fish.

The greatest depth of this lake is about twenty feet. Its margin is most of it fringed with water grass and lilies. The bank on the west side is high and dry, and on the east side there are fine building sites.

These waters swarm with fishes, trout, perch, gar. sacalait, rock, bar, choupique, casburgo, buffalo, cat, blowing fish, soft-shell turtle, and sardines. Some of the trout are two and a half feet in length, and sacalait eighteen inches.

This lake is fed by springs of excellent water, that break out around its margin. There is a large boiling spring in the middle of the lake.

The nearest approach of the Teche to this lake is about 700 yards. Its surface is about six or eight feet above the level of the bayou. There is water power here sufficient to drive a sugar mill.

PARISH OF ST. MARTIN.

GENERAL DESCRIPTION.

The extreme length of the Parish of St, Martin is about twenty-four miles, and its extreme width is about the same. It contains about four hundred square miles of rich prairie, swamp lands, heavily timbered, and tillable lands, covered with the finest body of timber in the State, suitable for sugar wood, building purposes, cabinet work, wagons, plows and wooden ware. And this supply of timber is immense, as we will show under its proper heading.

The parish is bounded on the north by St. Lnndry, by Lafayette on the west, Iberia on the south, and Iberville on the east.

THE TECHE LANDS.

The Bayou Teche enters St. Martin at its junction with Bayou Fuselier, at Arnaudville, formerly called Leonville; and, meandering through the parish, enters the Parish of Iberia six miles below the town of St. Martinsville, near Lake Tasse, thirty-five miles from Arnaudville.

The tillable land from St. Martinsville east of the Teche, is eight miles in width, including all the land between this bayou and Catahoula Lake. At Arnaudville, the tillable land on the east side of the bayou is three miles in width. The average width of the tillable land on the east side of this bayou, in its entire course through the parish, is over five miles. And the average width of the tillable lands on the west side of the Teche, is over three miles. In places, in the great bends of the bayou, the width is much greater. In our estimation it is difficult to overrate either the beauty or the innate merit of this portion of Attakapas.

RICH SOIL.

As to richness of soil, and all the qualities that are essential in any soil—drainage, ease of cultivation, its wearing and enduring properties, in the production of sugar, cotton, rice, corn, tobacco, indigo, or any other crops now grown or ever grown in the same latitude—fruits, melons, potatoes, cabbages, turnips, and the whole list of field, garden, and orchard products, no portion of Louisiana can excel those of the valley of the Teche in the Parish of St. Martin.

FORESTS.

From the open prairie which runs parallel with and near the Teche, to the Atchafalaya, the eastern limits of St. Martin, it is almost an unbroken forest of the finest timber in Louisiana. In the swamps of the

Atchafalaya there are millions of cypress trees, tall, straight, and many of them from three to four feet in diameter. Between these swamps and the Teche prairie, on the tillable land, there is an immense, unbroken forest of ash, gum, hickory, black walnut, magnolia, live oak, white, red, and other oaks, linn, pecan, sycamore, and other wild growth of less importance. On the west side of the Teche, in the rear of the open prairie, extending from Bayou Fuselier and the Upper Vermilion, down Bayou Tortue to Lake Tasse, there is a forest of swamp cypress, and also of oak, ash, gum and other trees which grow on dry and tillable land. Both banks of the Teche are skirted with fine forests, which we will describe under another head.

THE VALE OF THE TECHE.

The lines of swelling forests in the rear take the place of hills in helping form the Valley of the Teche. This bayou, in its course through St. Martin, is extremely beautiful, in many respects more beautiful than the lower Teche as it meanders through St. Mary. Its first banks on both sides at St. Martinsville are nearly twenty feet high, at Breaux Bridge twenty-two feet, and at the junction nearly thirty feet above tide water.

The banks of the bayou have a slope of less than thirty degrees to the water's edge. Everywhere there are beautiful building sites along the bayou. The banks give the bayou, everywhere, the appearance of a huge canal. The water is not more than two and a half or three feet deep in summer and autumn, and the surface is but fifty or sixty feet wide; but for about six months in the year it is navigable for small steamers. One lock at St. Martinsville would render the bayou navigable to the Junction the year round.

In Yankeeland, the neighbors would have joined and made a lock in such a place twenty years ago, and as a consequence they would have added a million of dollars or more to the value of the lands of the parish, which would have been much more generally settled up many years ago. And they would have had at least a tri-weekly boat and mail between the Junction and New Iberia. It is astonishing that so important an enterprise has been so long neglected.

THE FORESTS OF THE TECHE.

The scenery all along, on both banks of the Teche, from St. Martinsville to the Junction, a distance of thirty miles, is the most charming and magnificent we have ever seen in any part of the United States.

The forest trees on both banks, the magnolia, ash, live oak, red, white, and other oaks, black walnut, linn, gum, pecan, hickory, sycamore, and other trees, are tall, graceful, and of generous growth. On thousands of acres the grass grows on a smooth surface under the waving branches of these noble trees These lands are far more beautiful than the famous woodland pastures of Kentucky; the trees have a more luxuriant growth, the foliage is richer, and hangs out on the broad branches in a more generous abundance, and the soil is rich beyond anything we ever saw in the Great West.

And it is the cleanest looking country we have ever traveled over. The beautiful smooth prairies look as though they had just been washed; the grass looks like a lawn neatly shaved by some "Fine Old English Gentleman," who prides himself on his aristocratic estate; the fat herds grazing upon these green prairies help in giving the finishing touch to this magnificent landscape scenery.

PRICE OF LANDS.

A beautiful place on this bayou, eligibly situated in all respects, three-quarters wood, five hundred acres in the tract, ten or fifteen miles above St. Martinsville, is offered for $3,000. A farm of 230 acres, plain house and improvements, fencing, etc., was sold last season for $1,650. A place of 100 arpents, improved, was sold for $2,200. A farm of 220 acres was sold at eleven dollars per acre, $2,420. A farm of 120 acres was sold for $3,000. One of 500 acres was sold for $3,500. One from 180 to 200 acres was sold for $4,125. We think all the heavy landholders in this parish will sell small tracts to suit purchasers at reasonable prices.

HEALTH.

The health of this country is undoubtedly excellent. There are no swamps. ponds, or wet prairie near the Teche to generate disease; the grown people and children look healthy, and there is a goodly number of old people in the parish. It is thought that one active young physician could do the entire practice between the Junction and St. Martinsville, a distance of nearly thirty miles.

FRUIT.

This is an excellent country for fruit. In passing up the Teche, to the Junction, by land, in August, we saw peach trees hanging full of fine fruit, raised with but little attention and trouble, the trees and fruit as healthy as any we have ever seen. All the fruits of this latitude thrive admirably in this parish.

POULTRY.

This country abounds in poultry. Some of the merchants buy eggs at five cents a dozen, and pay in trade, but the usual price is ten cents. A spring chicken is exchanged for a pound of sugar, and grown chickens never sold for more than twenty-five cents apiece.

THREE FINE ESTATES OF GENERAL DECLOUET, MR. LASTRAPES AND DR. WILKINS.

ESTATE OF GENERAL DECLOUET.

This estate of ten thousand acres or more, is situated on the east side of the Teche, in the southern portion of the parish. The mansion of Gen. Declouet is on the east bank of the Teche, three miles above St. Martinsville. It is a noble, two-story dwelling, with spacious halls, galleries, and passage, a dining room and table large enough to seat the patrons of a large hotel, high ceiling, large windows and doors, an airy, convenient and pleasant mansion of a large-hearted and whole-souled Creole gentleman "all of the olden time." His parlor, sitting room and halls are ornamented with portraits of his paternal and maternal ancestors for several generations, and other valuable paintings.

Gen. Declouet keeps up, more than any gentleman we have visited, the best original style of Creole living; a bountiful table, an abundance of everything, excellent servants, a cordial and easy welcome to strangers, and a delicate attention to all their wants and comforts. He has the happy faculty, almost universal among the original Creole planters, of making even the greatest strangers feel at their ease, and perfectly at home. The manners and hospitality of the original Creole gentleman are inimitable, and universally admired.

The walks, groves, gardens and outbuildings are all laid out and put up on the liberal plan of the ancient rich and generous Creole gentleman of this country. The dwelling is situated between the main road and the bayou.

The yards fronting on the Bayou Teche, and towards the main road on the prairie side, are equally spacious. A broad avenue sixty or seventy feet in width, and three or four times that length, with tall and beautiful China trees on either side, extends nearly to the bayou. A large surface is shaded on the bayou front. And twenty or more pines, forty or fifty feet high, with noble trunks and extended branches, make a fine avenue and shade on the other front. And fine live oaks, magnolias, figs, and other trees, covered with rich and abundant foliage, are growing at intervals in the rear of the pines and Chinas on either side.

The sugar house is spacious and convenient, and the engine and sugar mill are ponderous and complete. The sugar house is large enough for saving a crop of six or eight hundred hogsheads of sugar.

The General has been a large stock raiser, besides a cotton and sugar planter. He owns over ten thousand acres of the finest land in Attakapas, most of it heavily timbered, and rich, tillable land.

ESTATE OF MR. LASTRAPES.

The Lastrapes estate is situated on the east bank of the Teche, below the Junction. The dwelling and its surroundings, sugar house, cotton gin, stables, corn houses and numerous outbuildings, are on the east bank of the bayou, about seven miles below the Junction.

This is one of the oldest Creole families of the Attakapas, and, like the Declouet family, they keep up the generous hospitalities and customs of the wealthy and educated Creole of the ancient and happy days of Attakapas.

Mr. Alfred Lastrapes' mother still lives on the old estate. She is a venerable lady, 80 years old, of great dignity of appearance, stands as straight as a young lady of 18, and is extremely active. She must have been very handsome when young, for she is a handsome old lady. Her mind is still fresh and vigorous. Like Gen. Declouet's, the parlors of the Lastrapes family, are ornamented with ancestral portraits in a fine state of preservation. These portraits have been admired by three generations, reaching through a period of nearly a hundred years.

The Lastrapes family cultivate cotton and sugar cane, and have a large stock of cattle. They own over ten thousand acres of choice farming land of great intrinsic value, though its present market value is like most of the lands of this parish, extremely low.

ESTATE OF DR. WILKINS.

Dr. Wilkins is a Virginian by birth, brother of John D. Wilkins, deceased, formerly a planter of St. Mary, and well known to many of our citizens.

His estate is on Bayou Fuselier, half a mile from Arnaudville; near, or on the line which divides St. Landry and St. Martin.

His dwelling is large, airy and comfortable, and he is surrounded with all the real comforts and blessings of life. He has substantial and beautiful shade trees, and a fine garden. There is an air of comfort and independence in all of his surroundings. This excellent family give strangers a genuine old Virginia welcome. If the old style Louisiana Creole gentleman ever finds a successful competitor in the grace, kindness, ease and perfection of his native hospitality, it is in the family of an old style Virginian, and nowhere else.

We have given this brief reference to three leading estates and families in St. Martin, not so much on a complimentary account, as to show strangers that this country is not the rude and barbarous region that many very well informed people at the North and West suppose it to be.

IN THE WILDERNESS.

Though just beyond the limits of St. Martin, it may not be amiss to notice Arnaudville and its surroundings.

Arnaudville is a small village made of two or three stores, a postoffice, dwellings, etc., on a bluff at the junction of the Teche and Fuselier, about thirty-five feet above low water mark. The banks of both bayous are here very steep, and the scenery is wild and interesting. Here are plenty of fish and game. Forest trees hang over the banks of the bayous, and in places lock limbs and branches. The road leading north towards Opelousas, passes miles through one of the most enchanting forests we have ever seen in the South, or anywhere else. We passed over this road in August, 1869.

Queenly magnolias with their wealth of green glossy leaves, and large white flowers, noble oaks, and pecans, ash, gum, hickory, black walnut, and numerous other trees of rare beauty spread their friendly branches above us as we rode over a good road through this region of indescribable beauty.

To give our readers a better idea of the stately character of these forest trees, we will give a few figures. A large oak had been torn up by the roots, and lay across the road, except a section sawed out and removed to let carriages pass. It was solid, five feet and a half in diameter at the chopping place. Twenty feet from the roots, it was four feet in diameter; forty-eight feet, at the first limb, it was three feet in diameter; to the second limb, fifty-eight feet. A log could have been made of it sixty feet long, nearly three feet in diameter, at the small end, and five and a half feet at the large end.

BREAUX BRIDGE.

This is a pretty place made up of a few stores and houses. It is by water about half way between St. Martinsville and the Junction. It looks new, brisk and lively. It has a fine elevated site, and at no distant day will doubtless become a large and prosperous village. It has a livery stable, hotel, Catholic Church, good schools, a lyceum, a brass band, and other proofs of civilization and progress. The new bridge across the Teche at this place is high and well built, and is creditable to the place and parish.

The people are hopeful and resolute, and say they intend to make that place the great business depot, even if the Chattanooga Railroad crosses the Teche two miles below them.

LAKE MARTIN.

On the back part of this place, about two miles from the bayou, is Lake Martin, over a mile in length, and little less than a mile wide. The soil around is firm, and one may ride to its bank on all sides. Tall cypress and ash trees grow in the edge of the lake, but gum, oak, elm, and the small growth of the Teche make a magnificent shade, in hot weather, all around it. The soil is firm and well set in grass under the trees. Parties from Breaux Bridge, accompanied by the brass band, come here to enjoy themselves in the summer and autumn. The lake is full of fishes of various kinds, and here is the finest place in the South for picnics, fishermen and duck hunters.

GRANDE POINTE.

Grand Pointe is situated above Beaux Bridge on the east bank of the bayou, and has a front on the bayou ten miles in extent, from Breaux Bridge to the lower line of the Lastrapes plantation. The settlement extends back five or six miles from the bayou.

This settlement includes about one hundred and fifty Creole families. They live in islands of timber, and coves of prairie, and cultivate cotton, tobacco, corn, and some cane. The country is beautiful and the land is rich. The people live in small, cheap houses. Many of them are very industrious, and they are generally out of debt.

They keep up the old Creole custom of having neighborhood balls every Saturday night. The balls are generally made up of the sons and daughters of small Creole farmers who work all day and dance at night. There are not less than sixty fiddles and fiddlers in this settlement. They are a merry people; and those who think the young ladies don't know how to arrange their hair, and primp for the ball room, so as to make themselves look attractive to the beaux, are simply mistaken.

These Creole families are very friendly with each other, and are troubled but little with the jealousies, and quarrels which sometimes afflict neighborhoods. They are extremely sociable, and obtain a large amount of social enjoyment at a small expense.

The timber in Grande Pointe, four miles back from the Teche, is gum, pecan, oak, ash, elm, and hackberry. The lands here look dry. Coules and ravines run to the lower lands in the timber further back. The whole sheet of country drains well, and some of these ridges several miles back from the bayou were out of water in the overflow of 1867.

NATURAL DITCHES.

The sheet of prairie on either side of the Teche is everywhere grooved with ravines which extend back many miles to the bayous, swamps and lakes. They are nature's ditches, and take off all the surplus water that falls in showers, or long continued rains.

GAME AND FISHES.

In the amount of game and and fishes, this excels most of the other parishes excepting Vermilion. Hunters and fishermen, and fishing parties may have rare sport in this parish at any season of the year.

CROPS.

More attention has been given to cotton than cane, in this parish; but both crops flourish well, and the yield is remunerative and satisfactory. The soil is everywhere fresh, rich, and easy of cultivation.

ST. MARTIN, 1850 AND 1859.

In the year 1850, before the parish was divided, it contained 220 farms, and 940 dwellings, population, 4,740 whites; free colored, 531; slaves, 5,835; total, 11,107 square miles.

In 1859 the cultivated lands of St. Martin were as follows :

Cultivated in sugar cane, 8342 acres; in cotton, 7187 acres; in corn, 11,488 acres.

State tax in 1859, $14,984; parish tax, $9447; State and parish taxes, $24,431.

LARGEST CROP OF SUGAR.

In 1858 the sugar crop of St. Martin amounted to 13,548 hogsheads; molasses, same year, about 20,000 barrels.

ITEMS—PARISH OF ST MARTIN.

Corn is often sold in St. Martin at thirty-five to fifty cents a bushel. Sweet potatoes usually sell at seventy-five cents a barrel. No soil in Louisiana yields more sweet or Irish potatoes to the acre than that of St. Martin.

Mr, Ervin, who has a place on the bayou below Breaux Bridge, has eleven pecan trees. One year he lost two-thirds of his crop, and sold the balance for one hundred and seventy dollars. His best prices were thirty-five dollars a barrel. The trees bear well every two years.

Between St. Martinsville and the Junction there is land enough for 2000 fine farms of one hundred acres each.

The prairies and woodlands of St. Martin, the distant swamps excepted, are high and airy, and no country can be more healthy than this.

Gen. Declouet is selling twenty-acre lots of excellent land at twenty-five dollars an acre.

They make carrote tobacco at Grande Pointe, in this parish. They sometimes make three crops in a year, two crops from suckers after the first crop is saved.

Mr. Alfred Lastrapes, who owns over 10,000 acres of fine land in this parish, offers to sell lands in quantities to suit purchasers, at reasonable rates.

Prairie Gros Chevreuil extends from the lower line of the Lastrapes plantation to Arnaudville, nine miles.

The rise in the Teche above Breaux Bridge in the overflow of 1867, was thirteen feet and six inches.

Bayou Fuselier connects with Bayou Vermillion. By cleaning out these bayous. small steamers could reach Vermillion Bay from Bayou Teche, by this route.

By deepening a ravine a mile and a half in length, navigation could be opened between the Courtableau and the Teche. Steamers have passed through this ravine in high water, from the Courtableau to the Teche.

PARISH OF LAFAYETTE.

GENERAL DESCRIPTION.

Lafayette is the smallest of the Attakapas parishes. Its extreme length is about nineteen miles, its width about the same. Its northeastern boundary, made by the bayous Carancro and Tortue, is irregular; the other three lines are nearly straight.

This parish has an area of about 300 square miles, about one-eighth of which is swamp and timbered land; the rest prairie. According to the assessment roll of 1868, the cultivated land was classed as follows:

Land cultivated in corn, 10,944 acres; cultivated in cotton, 8217; cultivated in cane, 175; cultivated in rice, 106: other crops, potatoes, pease, etc., not mentioned.

SOIL—ITS GENERAL CHARACTER.

The soil of Lafayette parish is light, loamy, mixed with sand. It is generally about twelve inches deep. It rests on a clay subsoil which is rich in plant food, like the subsoil of all the other parishes in this section of Louisiana. The fertile properties of the subsoil are only developed by exposure to the sun, and mixing with the surface soil.

Fields in Lafayette which have been in cultivation for seventy years principally in corn and cotton, are still fertile.

By plowing in a crop of cow pease occasionally, the richness of this soil would be perpetual.

Two ordinary creole horses, or mules, can break up the soil, and one horse can cultivate it.

A good field hand can cultivate thirty acres of corn, or ten acres of cane, or fifteen acres of cotton.

PRICE OF LANDS.

The best unimproved lands can be bought in Lafayette for $10 an acre. Cheapest unimproved, $1 25. Improved lands sell at from $20 to $30 an acre.

Lands in this parish rent at from $2 50 to $5 00 an acre.

POPULATION, FARMS, ETC.

There were 1587 registered voters in the parish at the last registration, more than half of them white. The vote in the year 1860 was from 800 to 900. The population of the parish, white and black, is about 8000, about equally divided.

In 1850, there were in this parish, 441 farms, 630 dwellings; population, white. 3390; free colored, 160; slaves, 2170; total, 6720.

BEAU BASSIN.

The road leading from Vermillionville to Grand Coteau, runs through a beautiful agricultural region called Beau Bassin. It is twelve miles from Vermillionville to the Carancro Crossing, and about four miles from the road to the eastern boundary of Beau Bassin, which is the eastern boundary of the parish. The lands near Vermillionville are nearly level, but extremely productive. A few miles north, between the road and the bayous, the surface becomes beautifully rolling.

The gentle slopes, and long, tortuous ravines may be ranked with the most delightful landscape scenery in Attakapas. Here we fiud some of the most pleasant building sites in this enchanting country. The swells are like the heaving bosom of the ocean after a storm. Descending into the ravines one feels as though he were in a trough of the sea, soon to rise up again on the mountain wave, and look out on the green ocean. The cottages of the farmers are many of them neat and comfortable. The green pastures, fat cattle, and fine fields of cotton and corn, in their proper season, indicate a rich soil, and a prosperous population. Shade trees and clumps of timber add greatly to the beauty of the scenery. The fields are generally enclosed with nice fencing of cypress pieux and posts, and the lands were formerly pretty well ditched. The country is airy and pleasant, and it is extremely healthful, as will be shown in our health statistics of the parish.

COTE GELEE HILLS.

The road leading from Vermillionville to New Iberia, after crossing the Vermillion river, two miles from town, enters a section of the parish of Lafayette called Cote Gelée, or the Cote Gelée Hills. This includes a division of country ten or twelve miles in diameter, southwest of Bayou Tortue, east of Vermillion, and on both sides of the main road to New Iberia.

The soil is here rich, the country beautifully undulating, with deeper ravines and higher swells than we find in Beau Basin. The farmers are thrifty, but not as independent as they are on the north of Vermillionville. Plain dwelling houses, and groves of China trees may be seen in all directions. The scenery in places is quite picturesque.

This is an open and airy country, with pleasant locations for residences, admirably drained, the soil rich, mixed with enough sand and vegetable loam to make it easy of cultivation. No portion of the South can be more healthful than this. Still, we would not wonder to learn that they have a full share of sickness here. They nearly all live in small houses, and have close shutters, and they sleep in rooms that are not well ventilated. They close their doors and shutters at night, and shut out the pure air so agreeable in well ventilated houses. Besides, they drink well water. They have but few cisterns.

They obtain pieux and posts for fencing in this country nearly as cheap as they can obtain them on the Teche, but firewood is scarce, and is obtained by trimming their China trees, and from the woodlands that skirt the neighboring bayous. Upon the whole, nature has done wonders for this country. A man who has twenty acres of land here should never be short of provisions for his family.

The range for cattle is admirable. A farmer should always have plenty of beef, mutton, pork, butter. poultry, etc., in such a country. And the people here are really independent. They owe little or nothing, and are thriving finely.

ROYVILLE.

This is a neat little village, but a few years old, in the open prairie, about 10 miles from Vermilionville, and south of Cote Gelée. Ii has a new and pretty Catholic church, 3 stores, a blacksmith and carriage shop, a doctor's office, and several dwellings, all new. It has a fine site for a village, and is surrounded by a thrifty population, with splendid prospects ahead. These prairies will, before many years, be settled up with small and industrious farmers.

SOUTH OF ROYVILLE.

There is a fine sheet of level prairie around Royville, and on towards Lake Peigneur. Many stock raisers and small farmers live in this prairie. Some of them are quite independent.

The Parish of Lafayette has many fine features, which those who merely pass through it on the main stage route cannot properly understand or appreciate.

VERMILION RIVER.

Darby, in his Geographical Description of Louisiana, thus speaks of this river :

The two vast prairies known by the names of the Opelousas and Attakapas, extend themselves on each side of the Vermilion, through its whole traverse, from its entrance into Attakapas to its egress into the Gulf of Mexico, the distance of one hundred miles.

Wood is much more abundant on the Vermilion than along the west bank of the Teche, and though the soil may be inferior in fertility, it is, nevertheless, excellent, and the quantity greater on an equal extent of river.

There are certainly eighty miles of the banks of the Vermilion, which have an extension backwards two miles, which affords three hundred and twenty superficial miles, or two hundred and four thousand eight hundred acres.

Some of the most beautiful settlements yet made in Attakapas are upon this river. From the diversity in soil and elevation, there is no risk in giving the preference in beauty of appearance to the banks of the Vermilion over any other river in Louisiana south of Bayou Bœuf.

If situations favorable to health, united to the most agreeable prospects, which are bounded but by the horizon, should be sought after; were taste to select sites for buildings, its research would here be requited, and be gratified by the breezes which come direct from the Gulf of Mexico. Fancy itself could not form a more delightful range than the Carancro and Cote Gelée settlements.

On leaving the dead level of the Teche, or the almost flat extension of the Opelousas Prairie, the eye is perfectly enchanted.

If a bold extent of view can give vigor to the imagination; if the increase of the power of intellect bear any proportion to the sweep of the eye, upon one of those eminences ought a seat of learning to be established. There the youthful valetudinarian of the North, would, in the warm, soft, and vivifying air of the South, find his health restored and his soul enlarged. Astonishing as it may sound to many, I do not hesitate to pronounce this, togéther with the range of hills from Opelousas, as the most healthy and agreeable near the alluvial land of Louisiana.

RESIDENCE OF MRS. CADE.

From the editorial correspondence published in the BANNER, May 18th, 1870, we copy the following description:

The residence of Mrs. Cade is situated one mile west of Royville and ten miles from Vermilionville. The dwelling is large and commodious, and comes square up to the first class planters' houses on Bayou Teche. It is situated in the midst of the most beautiful prairie in this country. The view from the upper gallery is perfectly enchanting. The limits of the semi-circular view in front, to the right and to the left, extend twelve or fourteen miles from the point of observation. Miller's Island, Isle Piquant, Cote Gelee Hills, the Teche woods above St. Martinville, are in sight; and islands of timber dot the beautiful prairie, in all directions, and the thrifty new village, Royville, is close by, its church steeple rising above the other white buildings. About three miles to the west the Vermilion woods may be seen, a fine sheet of prairie intervening.

Mrs. Cade has a large stock of cattle in Texas, nearly twenty thousand head. She brands about four thousand five hundred calves yearly, and has about three thousand beeves for sale.

At Mrs. Cade's we met Mr. Marsh, her father, who is now nearly eighty years old, and very active and healthy. Mrs. Marsh is the same age, and usually healthy. They have lived in this country nearly fifty years.

CHURCHES AND SCHOOLS.

There are three Catholic churches, two white Methodist churches, and one colored Methodist church in this parish. There are six church edifices. There are eight public schools in the parish.

POPULATION—HEALTH.

The white and colored population of Lafayette Parish is 8925; white and colored vote, 1785; white population, about 4000.

The ages of 116 of the oldest, native born, white male citizens of Lafayette, are as follows: Ages—60 years, 4; 61 years, 7; 62 years, 4; 63 years, 21; 64 years, 4; 65 years, 10; 66 years, 6; 67 years, 5; 68 years, 8; 69 years, 5; 70 years, 8; 71 years, 6; 72 years, 3; 73 years, 10; 74 years, 1; 75 years, 2; 76 years, 1; 77 years, 1; 78 years, 5; 79 years, 1; 86 years, 1; 88 years, 1; 93 years, 1; 115 years, 1. Total, 116.

Besides the above list, we have the names of two born in England, 63 years, and 72 years; one in Massachusetts, 90 years, and five in France, one, 67; one, 68; one, 69; one, 93, and one 98 years old.

As the women of Attakapas are more healthy than the men, the true health statistics of the parish must show that there are more than 116 white women in Lafayette of the ages given in the foregoing list. This would make over two hundred and thirty white men and women, natives of the Parish of Lafayette, in a white population of about 4000, whose ages range from 60 to 115 years, or one sixteenth of the population. Were those who are not native born added to the list, and also those who died from privations in the last war, the health statistics would show to still greater advantage.

There are thirty-two veterans of the war of 1812 in the parish.

Four active physicians could do the entire practice of the Parish of Lafayette.

After due inquiry, we have not been able to hear of the death of any young lady in Lafayette since the war, from the spring of 1865 to 1869. Deaths among children and young people in this country are extremely rare, but remarkably so among young ladies.

WHAT ONE FIELD HAND CAN DO.

Last year, on Mr. Patin's place, near Vermilionville, a young Canadian cultivated fourteen arpents, and made eight and a half bales of cotton and four hundred and fifty bushels of corn. He only paid $17 for help, and that was in cotton picking.

Two negroes, on the same place, made each about five bales of cotton, and about three hundred bushels of corn.

François Como made nine bales of cotton, and plenty of corn. He is a native of this parish. His father-in-law worked two negro men, and some small negro boys, about twelve years old, and made thirteen bales.

Two Spaniards, last year, made fourteen bales of cotton in this parish.

Three men cultivated eighty arpents of land with four Creole mules and two Creole ponies.

Two negroes made between nine and ten bales of cotton, and 400 barrels of corn.

Nearly half of the field labor of the parish of Lafayette is now performed by white men. About one-eighth was performed by white men before the war.

CROPS AND FRUITS.

Cotton, sugar, corn, rice, and all the field and garden crops of the other Attakapas parishes do well here. Corn, Irish and sweet potatoes, melons, peaches, pumpkins and field peas find a remarkably congenial soil here. All the fruits of the other Attakapas parishes, except oranges and the more delicate kinds, thrive finely in Lafayette. Formerly indigo was profitably cultivated here.

POULTRY.

This is one of the best parishes in the State for all kinds of domestic fowls. Some families are nearly supported by raising poultry. Domestic fowls are extremely healthy here.

PRICES.

In this parish fat beeves, four years old and upwards, sell at from $18 to $20; milch cows, $25 to $30; beef, per pound, in the market, 6 to 8 cents; mutton, 10 cents; corn, 50 cents a bushel. Sweet potatoes, 35 cents a bushel, or $1 a barrel; lumber, cypress, a thousand feet, $30 to $35; pieux and posts, $100 a thousand; bricks, $10 to $12; wood, $5 a cord; chickens, $3 a dozen; eggs, 10 to 15 cents a dozen; male turkeys, $18 a dozen; hen turkeys, $9 to $12.

FREIGHTS AND TRAVEL—DISTANCES—PRICES.

Freights sometimes are received and sent off by way of the Vermilion river, through Vermilion, Cote Blanche, St. Bernard (Bayou Sale), and Atchafalaya Bays, and the Atchafalaya river, Brashear City, and the Morgan L. & T. Railroad. Generally the freights are carted across the country between Vermilionville and New Iberia, and take the line of the Teche steamers and the Morgan railroad.

Carting between New Iberia and Vermilionville—twenty-five miles—dry barrels, 75 cents to $1 25; and from 15 to 20 cents a cubic foot. Wet barrels, $1 50 to $2; sacks of coffee and salt, 75 cents to $1.

Passage between Vermilionville and New Orleans, $10; time, 30 hours. Distance, via New Iberia, 167 miles; via Vermilion river and the Bays, and the Atchafalaya, 188 miles; via surveyed route, Chattanooga Railroad, 125 miles; Morgan Louisiana and Texas Railroad, 141 miles.

WILD GAME.

Wild game in Lafayette is usually abundant. Ducks, and some other game, are only found in the fall and winter. The following is the list found in this parish: Wood cock, partridge, prairie chickens, papabot, snipe, wild geese, canvasback, mallard, black and teal ducks, brent, robbins, rice birds, oppossums, racoons, rabbits, deer, and squirrels.

FISHES.

The fishes in the waters of this parish are the casburgc, perch, cat, buffalo, choupique, trout, sacalait, bar and gar.

GENERAL FACTS AND ITEMS.

The Bayou or River Vermilion is navigable 15 miles above the bridge on the New Iberia road, and 75 miles below the bridge to Vermilion Bay.

The largest crop of cotton ever made in this parish is 6000 bales. The largest crop of sugar, 2500 hogsheads.

Aggregate taxes of Lafayette: In 1868, State tax, $6940; parish tax, $1500.

The cattle, horses, mules and hogs are generally healthy. No charbon, blind staggers, murrain, or hog cholera. A new disease killed a few horses and cattle in 1869; the real character of the disease is not generally known.

Bees thrive well in this parish.

The China and Catalpa grow rapidly, and make excellent fire wood. Some families obtain nearly all their fire wood from the yearly trimmings of their China groves. Close pruning does not injure these trees. Limbs grow out ten feet long, and as large as a man's arm in a single year. They are easily propagated from seeds.

The best yield of corn per acre in Lafayette is about 60 bushels. The best yield of sweet potatoes, about three hundred bushels.

Formerly not half a dozen farmers or planters in Lafayette bought Western mules or horses. Creole mules and horses were almost exclusively used in cultivating the crops of the parish.

THE PARISH OF VERMILION.

GENERAL DESCRIPTION.

The parish of Vermilion contains about 1,600 square miles of land and water within its limits. About six hundred square miles of this is tillable woodland, prairie, and cypress swamp. About 500 square miles would include the prairie, and 100 miles the timbered land, the smaller portion of which is cypress swamp. Lakes, bays, and sea marsh cover about 1000 square miles of the surface of the parish.

About a quarter of the tillable land is on the east side of Vermilion river, or bayou, and three-quarters on the west side, extending to Lake Arthur and the Mermentau river.

The timbered land is principally on the Vermilion river, extending on both banks from the Lafayette line nearly to Vermilion Bay. The timber is narrow above Abbeville, but it becomes broad below this village, extending out a mile and a half on each side, in places. As it approaches the Bay it becomes narrower.

Below Abbeville there is a creek on the west of the river lined with a heavy body of timber, and there is another on the east side, the line of forest trees extending across the New Iberia and Abbeville road beyond the head of the creek.

There is a line of cypress timber on land a little higher than the prairie at the edge of the sea marsh north of marsh Lake, twelve miles long, and three-quarters of a mile wide. And there are islands of timber in the edge of the sea marsh east of Vermilion river.

There is also timber on the south side of Bayou Queue, Tortue, and on Pecan Island, and Grand Chernier.

SOIL AND SCENERY.

The soil of this parish is a dark vegetable mold, with a large proportion of sand, from eight to twelve inches deep; this rests on a subsoil of greyish clay, and beneath this is found the common yellow stiff clay. The soil along the Vermilion river has a larger proportion of sand than that farther back; this gives the soil a lighter color. On account of the larger proportion of sand here than in the Teche lands, these fields are easy to cultivate; and the roads need but little working, in most instances none at all, to make them good the year round. The bottoms of the ponds and ditches are not boggy. One may pass over any of them on horseback without inconvenience to the horse.

There are natural ponds in all these prairies where the stock cattle are supplied with water. These ponds are from 20 to 50 yards in diameter.

Prairie Gregg, which lies next to the sea marsh, southeast of Abbeville, is a beautiful sheet of land, level and rich, the soil darker than that east of Abbeville. The Gulf breezes sweep over it uninterrupted by forest trees. There are but few settlers here, and they are usually too poor to make any improvements of any importance, or to cultivate the land to any extent.

The prairie west of Vermilion river is described by a gentleman who lives in Abbeville, from an elevated position at the southern bend of the Queue Tortue, half way between the Vermilion and Lake Arthur, in the following language:

"The scenery here is the most perfect of its kind that the fancy can descry. Facing the south, one may here turn to the right or to the left, and as far as the eye can reach there is one vast expanse of natural meadow. Here and there may be seen a herd of cattle or horses almost hidden in some places by the tall natural grass. The prairie east, west, and south is dotted with little groves of trees, which shade the cheap and humble homes of the resident population, who live principally by farming, hunting and stock raising."

FOREST TREES.

The dry land timber is oak, ash, magnolia, gum, hickory, elm, beech, hackberry, and the usual dry land timber of Attakapas. The swamp growth is principally cypress.

CROPS.

The soil is good for sugar, cotton, rice, corn, potatoes, and all other products of the Attakapas parishes. The yield of cotton is not as large per acre as in higher latitudes; the parish is peculiarly adapted to the cultivation of rice. The largest yield of sugar per acre in this parish ever known is 4000 pounds, and this only on a few acres under highly favorable circumstances. A hogshead of 1200 pounds is a fair yield. The largest sugar crop of the parish, 2000 hogsheads; the largest cotton crop, 3000 bales. The yield of corn per acre is from 25 to 50 bushels; rice, 1200 pounds of clean rice; sweet and Irish potatoes do as well in this as in any other part of the State. The capacity of the soil has never been well understood on account of the general attention to stock raising at the expense of the interests of agriculture.

FRUITS.

No part of Attakapas can succeed better in raising peaches than in this parish. The soil on the banks of most of the bayous, and in much of the prairie, is admirably adapted to this fruit. And the general appearance of the peach trees justifies this conclusion. Oranges and the mespilus do well in the lower part of the parish. All fruits that do well in Iberia, St. Martin, Lafayette, and St. Landry do well here, both cultivated and wild fruits.

PRICE OF LAND.

Lands for sugar planting, an abundance of wood, improved or unimproved, can be purchased at from $10 to $30 an acre. Lands for cotton planting, without much wood, can be purchased at from $3 to $10 an acre. There is an immense amount of government land in the parish suitable for rice or stock farms, which may be entered at government prices.

TEAMS FOR CULTIVATING.

Oxen are generally used in breaking up new ground, and creole horses in cultivating it. Oxen are not put to work till the grass rises in March, since few of them are fed on hay, or corn. Agriculture here has received less attention than in the other parishes.

WELLS.

Good well water can be had at the depth of about 20 feet. It is cool, and usually pronounced good and healthful water.

POULTRY.

Chickens and eggs to the value of $20,000 and more are sent from this parish yearly to the New Orleans market. The peddlers buy up more than $5000 worth of poultry and eggs as they travel from house to house on the praries and bayous. Eggs often sell for ten cents a dozen, and spring and grown chickens from ten to twenty-five cents. Wild ducks often sell two for twenty-five cents; wild geese sell for fifty cents.

WILD GAME.

There is more wild game in this than in the other parishes. Ducks abound here in their proper season, and geese and brent are also abundant. Prairie birds are abundant. The market here is often glutted with ducks and other wild game offered at low prices. There are large numbers of deer and wild hogs in Vermilion.

FISH AND OYSTERS.

Vermilion Bay abounds in fish and oysters, and the fresh water lakes, the ponds and bayous have an abundance of fish such as are common in other parishes.

PRICES.

Pieux sell for $100 a thousand; lumber, $30 to $35; corn $1 a bushel; sweet potatoes, twenty-five cents a bushel. Fat beef in the Abbeville market sells at six cents per pound. Wild hogs are brought from Pecan Island and sold cheap.

HEALTH.

There is not a more healthy spot in the South than this parish. The same diseases prevail here that usually prevail in the other Attakapas parishes. The scarcity of timber between the prairies of the parish and the Gulf allow the health giving sea breezes free circulation and free passage to nearly every man's house.

POOR MAN'S PARISH.

There is no parish where a man who is poor and industrious can more easily make a living than in the parish of Vermilion. The soil, climate, game, fish, oysters and beef are all his friends. With a few cows, a flock of sheep, poultry, a pair of mules, a piece of land, a gun, and common industry and economy; he ought to prosper.

CHURCHES AND SCHOOLS.

The prevailing religion in the parish is the Roman Catholic. There are some Methodists and other Protestants in the parish. Educational advantages are not as good as in the other parishes.

ABBEVILLE.

There is good navigation in the Vermilion River the year round the whole length of the parish. Abbeville is thirty-five miles from the mouth of the river. The population of this place numbers about 500. The trade of Abbeville amounts to about $150,000 a year. Vessels ply between this place and Brashear, and between this river and New Orleans the year round.

MARSH LAKE.

Marsh Lake, in Vermilion parish, is usually called White Lake. So seldom do the inhabitants penetrate this part of the marsh, that many people raised in the parish think there is no such lake to be found. It is a sort of unexplored region to the inhabitants.

ISLANDS.

Grand Chenier, in the southwestern portion of the parish, is about twelve miles in length, and contains a thrifty population of intelligent farmers. The soil is rich, and tropical fruits, sugar cane, sea island cotton, tobacco, and all other products of Attakapas thrive well here.

Pecan Island, situated in the sea marsh, six miles from the sea coast, in the southern part of the parish, is sixteen miles in length. It is covered with live oak and pecan trees, and it sustains numerous hogs and cattle. There are several families living on it. They cultivate but little land, but obtain a handsome income from their hogs and cattle.

This island, supposed to have been the resort of the pirate Blue Beard and his men, and other more modern pirates, has often been visited by parties in search of hidden treasure. Trees have been chopped into in search of copper nails to get the bearings of pots of money, and pits have been dug when the bearings were agreed upon.

There are many human bones buried here, supposed to be the bones of prisoners brought here by the Attakapas Indians, who were cannibals, to be stewed into chowders with clams. They are said to have been very fond of the dish.

Cheniere au Tigre is in the southeastern part of the parish in the sea marsh, near the South West Pass of Vermillion Bay. This is a famous stock ranche. Here beeves, as in all other places near the coast, keep fat the year round and are ready for market in January and February. Not less than six thousand head of cattle live in the Marsh along this coast. Stock owners live in small groves of timber, and on slight elevations of land near the coast,

BETTER PROSPECTS.

Though the parish of Vermillion has been passed by or overlooked to a great extent, by the traveling public, it has great merits, and these merits will, before many years, be better understood.

Its situation away from all the great thoroughfares through Attakapas has been the principal cause of its not naving been more generally visited, and better known. Strangers who come to Attakapas will do well to visit Vermillion parish and decide upon its merits for themselves.

PARISH OF ST. MARY.

GENERAL DESCRIPTION.

The parish of St. Mary has a front on four great bays connected with the Gulf of Mexico, forty miles in extent It has an average width of a little. more than twenty miles. It is about fifty miles by the main road through the parish from its western line, near Jeannerette, to its eastern line at the Bœuf crossing of the Morgan Railroad.

St. Mary appears to splendid advantage from the pilot house of steamboat as she plows through these navigable bayous lakes and bays, and to poor advantage on the best map that can be drawn.

GENERAL ELEVATION.

The highest land in St, Mary, excepting the islands Cote Blanche and Belle Isle, is not over fifteen feet above the level of the Gulf of Mexico. There is a daily tide of from one to two feet in all of her bayous and lakes. The highest land around Berwick's Bay, has an elevation of about ten feet; and from the Bay to Pattersonville, and three or four miles up the mouth of the Teche, the elevation is but little above that around the Bay and on the Bœuf.

At Franklin the west bank of the Bayou Teche is but about thirteen feet above tide water, and the east bank is a little lower. Below Jeannerette the elevation is about fifteen feet. The two Islands, Belle Isle and Cote Blanche, at their highest points, rise more than one hundred and sixty feet above the level of the Gulf.

The sea marsh is most of it under water during storms from the Gulf sweeping towards the land.

THE SOIL.

There is not an acre of poor land in the parish. Fields that have been cultivated in corn and sugar cane for nearly a century, without manure, still produce good crops. The lands are easily and cheaply restored after long continued and severe cropping. The parish has land restoratives within its limits better than Peruvian guano, as we will show in an article under its proper heading.

AGRICULTURAL PRODUCTS.

Cotton is cultivated in St. Mary, but it is not considered a profitable crop. Sugar cane is the true crop of the parish. Much of the land of the parish is adapted to rice. The sea marsh, by local levees and draining machines, makes rich lands. This soil consists principally of a vegetable deposit of great depth. Swamp lands or any of the reclaimable wet lands, are fine for rice, corn, sweet and irish potatoes, pumpkins, pease, beans, indigo, ramie, arrow root, ginger, castor oil beans, tobacco, hay, cabbages, and turnips do well in this soil and climate, though a part of this list of articles has never been cultivated only to a limited extent. Sea Island cotton does well on the islands along the coast.

GARDENS.

Garden vegetables grow in this parish the year round. Nearly all kinds of vegetables grow the same here as in the North and West. The winter gardens cuntain onions, mustard, eschallots, leeks, garlic, beets, cabbage, carrots, turnips, cress, roquette, lettuce, radish, cauliflowers, celery, etc., etc. Good gardeners have an abundance of vegetables fresh from the garden the year round. White head cabbages, and fine ruta baga and red top turnips may be taken fresh from the garden in January and February, and also in the summer and fall.

CROPS, TRADE, ETC.

Thirteen thousand slaves were formerly owned in this parish, valued at about six millions of dollars. Before the war, about fifteen steamers were engaged on these bayous, lakes and bays, in the busy season of the year, and as many as one hundred and twenty-five vessels have cleared at the port of Franklin, for Northern and Southern ports, freighted with sugar, molasses and live oak, in one season.

The yield, per acre is, in an ordinary season, a hogshead of sugar and fifty or sixty gallons of molasses; in a good crop year double that amount. The sugar crop is cultivated nearly the same as corn. In boiling the crop it usually takes about three cords of wood to the hogshead. The crop is laid by before July, and sugar making commences in the latter part of October, or early in November.

RICE CROP.

A Louisiana rice planter gives the following statement:

"Rice lands well cultivated, not flooded, produce 6 barrels of 250 pounds each to the acre, or 1500 pounds. Flooded lands produce ten barrels. The flooded rice weighs heavier than rice not flooded. One hand with proper implements and teams, can make ten acres of unflooded rice, and more of flooded. In less than four months from the time the ground is plowed to receive the seed, the riee crop may be harvested.

Rice is cleaned at the rice mills at a cent a pound. There are twelve or fifteen rice mills now in operation in the State, and they are generally doing a good business. Most of them are situated in the parish of Plaquemines.

THE ORANGE CROP.

The yield of oranges, per acre, is enormous. It is impossible to make any estimate that is reliable, as we have not the acres or yield of any one orchard; but below New Orleans single orchards sometimes yield from $10,000 to $30,000 yearly at a dollar a hundred oranges. The largest orchards produce over three millions of oranges yearly. Some trees commence bearing when they are five or six years old, and earlier bearing can be produced by graftiug and budding.

A full grown, healthy orange tree, fifteen or twenty years old, in a good season, will produce five thousand oranges. It takes from three to four hundred oranges to fill a flour barrel. So the largest orange trees produce from forty to fifty bushels of fruit in a favorable season. Oranges usually sell, on the trees, at ten dollars a thousand.

TOBACCO.

The profits of tobacco culture in this country are satisfactory, but it takes too much skill and care to make and save a good article. Sugar and rice are less troublesome, and more profitable. Perique tobacco is generally produced in St. James parish, but it may be made in St. Mary. It is the best smoking tobacco in the world. Perique snuff is not excelled by any other.

FRUITS.

Fruits of various kinds ripen in St. Mary from April to November. The mespilus, or Japan plum tree, a beautiful evergreen, as large as the orange, blossoms in the fall, the fruit grows during the winter and ripens in March, except when the winters are uncommonly cold; then the fruit fails. The fruit is yellow, pear shaped, and about an inch in length, and very good.

Dewberries, large and abundant, grow wild all over the parish; the ripen in April.

Blackberries are abundant; they ripen in May.

Strawberries are prolific when properly cultivated, and continue in bearing six or eight weeks; they ripen in April and May.

Eight or ten kinds of plums ripen in June and July.

Eight or ten kinds of figs ripen in July and August.

Peaches ripen in July and Auguet.

Apples ripen in July, August and September.

The muscadine grape, or black scuppernong, grows wild on the banks of all our bayous, and in the forests. It ripens in August. The white scuppernong grape thrives finely, especially on the islands of the coast.

Pears of superior quality grow on the banks of the Teche, and thrive well; they ripen in August.

Olives do well in this parish, but no attention has been given to their cultivation. They would do well on Belle Isle, Cote Blanche, and the other islands.

Almonds do well in St. Mary; they ripen in the fall.

Pecans ripen in September.

Oranges ripen in October, and frequently hang on the trees till December, improving in sweetness all the while. This is the queen of fruit trees. Its robes of deep green in mid winter are beautiful, and its myriads of beautiful, white fragrant flowers in early spring are only eclipsed by its golden fruit in Autumn. Oranges are so plentiful in the lower part of the parish that they are frequently given away by the barrel, and seldom sell for more than a dollar a hundred. They are much finer than Cuba oranges.

Bananas, lemons, limes, and shaddocks ripen in October; these are more delicate than the orange tree, and seldom do well without a liitle extra protection, except in favorable locations in the lower part of the parish.

Pine apples may be raised in the parish with slight protection. Doubtless other valuable tropical fruits will be introduced into this country, and be ranked, in time, with the staple fruits of this parish.

Apples, currants, damsons, gooseberries. English cherries, and perhaps a few other northern fruits, do not thrive well in this climate.

It will be understood that we do not state that all the fruits in our list are found in abundance in St. Mary. We merely wish to state that experience has proved that they may be produced in abundance, excepting apples and a few other fruits, if the people will cultivate them. Oranges, plums and figs are the only cultivated fruits that are abundant, and these require but little care or culture.

This section of Louisiana is better adapted to fruit culture than any other portion of the United States. The fruit here is less troubled by worms, bugs, insects, and diseases than in any Northern State.

Fruit culture in St. Mary is yet in its infancy. When as much attention and skill are bestowed upon fruits there as have been brought to bear upon the same business in the Middle States and New England, our parish will be a paradise, all except the forbidden fruit.

SWAMPS AND TIMBER.

In the rear of nearly all the plantations in the parish there are cypress swamps, containing a heavy growth of trees, for building and fencing purposes, for making sugar hogsheads and molasses barrels, and all other purposes for which cypress lumber may be used.

CLIMATE.

Our parish is favored with a comfortable climate. Strangers from mountainous and hilly regions cannot understand how this can be, but we will submit a few facts on the subject.

This parish borders on the Gulf coast. We have healthful and cooling sea breezes during the summer and fall. Persons sleeping in rooms that are well ventilated never complain of hot or uncomfortable nights, not even in July and August. In traveling on our steamers on these waters in July and August, by night, seated in front, the air is sometimes almost too cool to be comfortable. In the summer of 1867, when the thermometer in New York and Philadelphia went up to 103 in the shade, in Franklin it did not go above 92.

The large surface of water, lakes, bays and bayous, around and within St. Mary, tempers the summer heat and winter cold, and the bland south breezes from the Gulf bring comfort, health and healing in their wings,

The first and lightest frosts seldom appear till in November. We have not the statistics of the weather in this locality, but those of a parish a little further south than this, show that in the last seventeen years the first frosts appeared three years in the latter part of October, eleven years in November, three years in December. Our winters here are merely the climate of Northern autums.

HEALTH.

Our climate is decidedly healthful. Chills and fever, and diarrhœa are the principal diseases, and these are generally brought on by imprudence or carelessness, and usually yield readily to remedies if applied promptly. Congestive chills are extremely rare. Common fevers and chills yield to the simplest remedies, with which every body is familiar. People seldom die either of fever or diarrhæa. Consumption is a rare complaint in this climate. Rheumatism and most other complaints of higher climates are rare in St. Mary.

No more healthy women can be found in any other part of the United States than are found in this parish and the parishes extending through to St. Landry. There were but two deaths among the young ladies of this parish for four years terminating in 1869, and one of these came to her death by a fall from her horse, and the other from yellow fever contracted outside of the parish.

The married ladies of this country are much more healthy than those of the North and West, and the men are generally healthy. Much of the sickness among the men is from imprudence, irregularities, intemperance, or from fevers or diarrhæas not seasonably prescribed for by a competent physician.

The yellow fever has never been an epidemic in Franklin, Centreville, or any village above Franklin, but twice since the country was settled. It is not a disease of this climate.

Nearly all that Dr. Tatman has said about the health of St. Landry, in the article on that parish, will apply to most all sections of the Attakapas parishes. But in St. Mary all portions of the parish are not equally healthy, and the same may be said of the other parishes inclnding St. Landry.

POPULATION.

Before the war the white population of the parish numbered about four thousand, and the largest vote ever cast was short of one thousand. Our people have always been noted for their hospitality and for their love of law and order. The majority of our people were decidedly opposed to secession, and were in favor of Bell. or Douglas. For this reason they were not included in the original emancipation proclamation of President Lincoln. But during the war nearly all of the citizens of the parish sided strongly with the South, and as soon as the war was over they ardently desired peace, and intended to act in good faith toward the old government and flag.

Northern gentlemen and families who have settled among us since the war, will testify that they have been treated kindly, and that they can live as securely here as anywhere in the North or West. The stranger and the freedman will be as fairly dealt with by a St. Mary jury as the original citizens of the parish.

PLANTATIONS AND FARMS.

There are about one hundred and fifty plantations and farms in St. Mary within its new limits, its western line extending from the upper line of the Grevemberg plantation, near Jeannerette, striking between Cypremort and Grand Cote, or Week's Island.

There are twenty plantations on Bayou Salé, nearly all of them in cultivation. Twenty-five years ago there were twenty-five sugar mills on that bayou all run by horse power. Bayou Salé is about twenty miles in length; the tillable land on it is about a mile wide; plenty of cypress in the rear of most of the plantations. These are the best sugar lands in the parish.

Bayou Cypremort has fifteen plantations; the tillable land is wider than that of Bayou Salé. It has an abundance of timber, ash, gum, oak, magnolia, and a considerable amount of cypress.

From Franklin to the mouth of the Teche, the distance is fifteen miles. On this part of the bayou there are thirty-six plantations. The depth of the tillable land on both sides of the bayou, on which these plantations front, is over two miles, in some places three.

The distance on the Atchafalaya from the mouth of the Teche to Berwick's Bay is about twelve miles, and on this part of the Atchafalaya there are about twenty-four plantations, some of them small. The tillable land is about the same width as that of the Teche.

On Berwick's Bay and the Bœuf there are not half a dozen plantations in running order. There are about twenty plantations and small farms on the bay and bayous, and lakes near it.

From Franklin to the upper line of the Parish of St. Mary there are, on the Bayou Teche, about forty plantations and farms.

The width of land on the Teche, above Franklin, is greater than that below, and the land is higher. In places on each side of the bayou the tillable land is more than three miles deep. South of Jeannerette, on the head of Bayou Cypremort, the land is from four to five miles in depth. There are more small farms in this parish than formerly. Nearly all of the plantations of the parish are now in cultivation.

PROFITS OF SMALL SUGAR FARMS.

Mr. E. Meynard, a native of Louisiana, and nearly raised in St. Mary, in 1868 rented a small farm at Charenton, nine miles from this place, planted and cultivated forty-five arpents of land, about 37½ acres. He planted in cane 22 arpents; corn, 15 arpents; rice, Irish and sweet potatoes, etc., 8 arpents.

He hired a white man to assist him in his crop six months, and paid him ninety dollars. He hired negroes occasionally for a few days. This cost him in all fifty-five dollars. His whole labor account for the season was but one hundred and forty-five dollars.

HIS CROP.

He made sugar cane enough to yield forty-five hogsheads of sugar, and sixty barrels of molasses.

He made rice enough to last his family two years.

He made Irish potatoes enough for his own use, and sold the surplus for one hundred and thirty-five dollars.

He made corn and fodder enough to supply his place for one year.

Not having a sugar mill of his own, he contracted to have his cane hauled and worked up four miles distant, and gave one-third of the crop as toll.

GROSS CASH PROCEEDS FOR MR. MEYNARD.

30 hogsheads of sugar at $120 a hogshead....................	$3,600
40 barrels of molasses at $20 a barrel.........................	800
Irish potatoes sold at..	135
Rice, sweet potatoes, and other products.....................	50
	$4,585
Whole cost of labor....................................	145
Proceeds of Mr. Meynard's labor................	$4,440

Mr. Brownson, who made up Mr. Meynar's crop, sold his toll, 15 hogsheads of sugar, and 20 barrels of molasses, for over $2500. Gross sale of sugar, molasses, rice and potatoes, from the entire crop, $7,085.

The above results are as true as they are extraordinary. The seed cane was good, the season good, and the industry and management could not have been excelled by any one. Mr. Meynard, at the close of the war, came home from the Confederate army without a dollar—he now owns the place on which he made the above crop.

HEDGES.

The pyricanth makes the best hedges in this country. It is propagated from cuttings, is an evergreen, beautiful, compact, full of short thorns, grows thick close to the ground, can be trained to any desired shape, and makes a good hedge in a few years. The Cherokee rose is worthless. The Chickasaw rose makes a good hedge, but it makes a mountain of vines and foliage. The bois d'arc makes a good hedge, but it requires too much labor and is too much inclined to grow tall, and to form trees.

THE CHINA, CATALPA AND BLACK LOCUST.

The China is a fine shade tree. Bugs or worms will not live on or around it. It is propagated readily from seeds, makes good fire wood, even when green, makes good cabinet wood, grows rapidly, not easy to decay, makes good fence posts. The limbs cut from trees planted near houses in the prairies supply many families with wood. Its growth is rapid, and it bears close trimming.

Nearly the same facts hold good in regard to the Catalpa and the black locust.

FERTILIZERS.

The deposits in the bottom of all the bayous of St. Mary are rich beds of muck, into which a pole may be run to the depth of ten feet or more. This is an excellent manure for gardens. Swamp muck is fine for gardens. The supply is inexhaustible. The sea marsh deposit is a fine fertilizer; but the best and cheapest of all is the cow pea. All sugar planters will agree to this fact. Planted among the corn, and the vines plowed in, the land becomes strong and productive at once. So the planter may get a full crop of corn, and enrich his land with a crop of cow pea vines the same year, at a trifling cost per acre.

OVERFLOWS.

The west bank of the Teche, from a point five or six miles below Centreville, to its source in St. Landry, has not been overflowed in the memory of man, and it has no levees to protect it. And this bank protects Bayous Sale, Cypremort, and all of the country west of this bayou.

The lands in the lower part of the parish, and on the east side of the Teche, were overflowed in 1778, 1828, and 1867. When Grand Levee, on the Mississippi, stands firm, no part of St. Mary can suffer by overflows.

GENERAL ITEMS.

The fishes of the waters in and around St. Mary, are red fish, black drum, trout, sheep head, flounder, mullet, croakers, cat, buffalo, perch, soft-shell turtle, gar, and choupique.

White men stand field labor in St. Mary as well as colored men, and they have less sickness and mortality than colored men.

Milch cows, when properly attended to, do well in this parish. No richer milk or finer butter is produced anywhere than that formerly produced from the best cows on Bayou Teche.

Hogs, chickens, and all kinds of poultry do well in this parish, excepting turkeys, which, of late years, have not done as well as formerly.

Steamers may land at nearly all of the plantations in this parish.

The parish is situated on tide water, and never suffers by freshets from heavy or long continued rains.

The crops of St. Mary are laid by, and field work stops, or may stop, by the first of July.

The Teche is considered the most beautiful bayou in the State.

BAYOU CYPREMORT.

We can give no adequate idea of the beauty of the forests of this bayou. Both banks are lined with tall, majestic magnolias, many of them fifty or sixty feet high, and clothed with a foliage which in beauty of hues and gracefulness of their garments, beggars description. Its millions of large, dark green leaves, the upper surface polished and glistening in the sun, the under side of a delicate brown color, the graceful form and noble bearing of the tree, and, in its season, its myriads of large, white fragrant flowers ornamenting all parts of its rich foliage, from its summit to its base, secures to it rightfully the title o queen of the forest.

Mingled with the magnolia on this bayou we everywhere find the elm, sweet gum, ash, oak, black walnut, pecan, hickory, and a rank growth of grapevines climbing the tall trees, and burying saplings and the small undergrowth beneath them, forming vegetable mounds as large as a dwelling house of medium size.

The road leading through these enchanted forests, along the banks of the bayou, is firm, smooth and sandy. The bayou itself is by no means beautiful, since it is usually filled with rank weeds, rushes, willows and numerous other trees and bushes peculiar to these shallow and narrow bayous. It can in no place be navigated with a skiff.

COTE BLANCHE.

This island is about ten miles from Franklin by water, and twenty by the buggy road via Cypremort and across the marsh.

Cote Blanche rises up an island mountain out of the marsh by the Gulf of Mexico. Its highest elevation is one hundred and eighty feet above the level of the Gulf. It has hills and dales, valleys and planes, lakes and springs, a rich soil, and a climate in which it is hard to get sick or die; the pure sea breezes from the Gulf fan and cool its surface during the summer and autumn months, and temper the winds of winter. On the south side, next to the Gulf, is a bold precipice a hundred feet high, whose base is washed by the salt waves. Here is fine bathing when the tide flows in; the beach is firm and smooth, and the firm bottom gradually deepens so that bathers may wade out a hundred yards.

On the bluff behind the precipice, overlooking the Gulf, the surface is rolling. It is firm, smooth, and sandy, a fine site for a village of pleasure seekers. And such sleeping and bathing as may here go almost hand in hand, few have ever enjoyed.

This island, at some future day, may be made one of the most beautiful spots on the face of the earth. Here sugar cane, Sea Island cotton, tobacco, rice, corn, sweet and Irish potatoes grow in the greatest luxuriance. And grass abounds where the plow or shade trees do not oppose it. And when we come to the fruits what may we not say of it? This island of two thousand acres may one day become almost one unbroken vineyard, and the best wines and brandies in large quantities may be exported from it. Here olives, oranges, lemons, bananas, citrons, limes, and many other tropical fruits may be made to bring large revenues to the island. The mespilus, peaches, figs, plums, dewberries, blackberries, strawberries, all do well in this favored spot. And here melons and garden vegetables grow and thrive as they seldom thrive elsewhere.

In addition to all its other merits, it affords the finest pastures for cattle and horses, a fine range for hogs and domestic fowls. In the waters in front, and the bayous around the island, the supply of fine fishes is inexhaustible, and oyster reefs in abundance, deer, geese, ducks and brent are at the service of epicures and hunters.

The broad sheet of marsh around this island furnishes the best of winter range for cattle. Thousands could find ample support as they do in the marsh in other portions of the Gulf Coast. After the first frosts of winter appear, the immense sheet of stubble is burnt off, and an abundant growth of young grass continues to spring up during the winter and spring. The most of the surface of this marsh is firm enough to bear up horned cattle as they rove over it for food, when the grass on the prairies is dead.

So much for Cote Blanche Island, now the property of William Fellowes, Esq., of New York.

SUGAR CROP OF ST. MARY, 1869.

	Hhds.
Bayou Teche—New Iberia to Franklin	14,155
Franklin to mouth of Teche	9,461
New Iberia to mouth of Teche, 60 miles	23,616
Atchafalaya—Mouth of Teche to Berwick's Bay, 12 miles	5,394
Berwick's Bay	1,818
Bayou Bœuf	3,317
Bayou Salé, 20 miles	3,957
Bayou Cypremort, 20 miles	2,443
Weeks' or Cote Blanche Island	711
Petit Anse, or Salt Island	662
Cypremort, Au Large, and Petite Anse Prairies, Grand Lake, etc., etc	2,716
Sugar crop of St. Mary, 1859	44,634

Molasses, same year, about 70,000 barrels, 40 gallons each—2,800,000 gallons.

SUGAR CROP OF ATTAKAPAS AND ST. LANDRY—1858-9.

PARISHES.	Hhds. sugar.	Bbls. molasses	Steam.	H'se Power.	TOTAL.
St. Mary	44,634	66,941	91	82	173
St. Martin	13,548	20,328	36	41	77
St. Landry	7,388	10,082	32	12	44
Lafayette	1,286	1,928	2	9	11
Vermilion	862	1,292	1	13	14
	67,718	100,571	162	157	319

GEOGRAPHICAL NAMES—LOCAL PRONUNCIATION.

Name.	*Pronunciation.*
Attakapas	At-tak-a-paw.
Atchafalaya	At-chaf-a-lyre.
Au Large	O Larzhe.
Brashear	Bra-sheer.
Bourbeux	Bore-bu.
Bœuf	Burf.
Bayou	By-you.
Brulé	Broo-la.
Bute a la Rose	Bute-ar-lar-Rose.
Boutté	Boo-ta.
Courtableau	Kore-tar-blo.
Cote Blanche	Kote-Blarnsh.
Coulée	Koo-la.
Cypremort	Sip-re-more.
Cote Geleé	Kote-zher-la.
Carancro	Kar-arn-kro.
Calcasieu	Karl-kar-shu.
Chicot	She-ko.
Coquille	Ko-kee.
Cheniere Au Tigre	Shen-eer-o-Teeg.
Charenton	Shar-arn-taun.
Fordoche	For-dosh.
Faquetaike	Fak-i-tike.
Fausse Pointe	Fawse Point.
Mermentau	Mair-men-tow.

Marmou	Mar-moo.
Nez Pique	Na-pe-ka.
Petit Anse	Pet-it-Awnse.
Portage	Por-tazhe.
Peigneur	Pain-yur.
Palourde	Pa-loord.
Queue Tortue	Ker-tor-too.
Shaffer	Sha-fur.
Salé	Sal-a.
Teche	Tash.
Tasse	Tas.

A Poet's Views of Attakapas and St. Landry.

PASSAGES FROM LONGFELLOW'S EVANGELINE.

That magnificent portion of Louisiana, west of the Mississippi, the Teche and Opelousas region, usually called "Attakapas and St. Landry," the land of enchanting scenery, of beautiful bayous, and glassy lakes, and bays; of splendid prairies, and noble forests; of pleasant skies, and gentle breezes; the land of flowers, of beauty, and of health, is the land where Evangeline sought her lover, Gabriel, the son of Basil, the blacksmith, as described in Longfellow's exquisite poem, "A Tale of Acadie."

In 1755, the village of Grand Pré, in the Province of Acadia, or Nova Scotia, was broken up by Gen. Winslow, under orders from the King of England, the property of the Acadians forfeited to the Crown, and 253 of their houses were set on fire at one time. Among others, Basil the Blacksmith, the father of Gabriel, had proceeded to Opelousas, in St. Landry, and Evangeline followed Gabriel, her lover, with her guide, the Father Felician. Passing down the Mississippi, as they approached bayou Plaquemine :—

"Level the landscape grew, and along the shores of the river
Shaded by China trees, in the midst of luxuriant gardens,

Stood houses of planters with negro cabins and dove cots.
They were approaching the region where reigns perpetual summer,
Where through the Golden Coast, and groves of orange and citron,
Sweeps with majestic curve the river away to the eastward.
They, too, swerved from their course, and entering the bayou of Plaquemine,
Soon were lost in a maze of sluggish and devious waters,
Which, like a network of steel, extended in every direction.
Over their heads the towering and tenebrous boughs of the cypress
Met in a dusky arch, and trailing mosses in mid air,
Waved like banners that hung on the walls of ancient cathedrals.
Deathlike the silence seemed and unbroken, save by the herons
Home to their roosts in the cypress trees returning at sunset,
Or by the owl, as he greeted the moon with demoniac laughter.
Lovely the moonlight was, as it glanced and gleamed on the water—
Gleamed on the columns of cypress and cedar sustaining the arches,
Down through those broken vaults it fell as through chinks in a ruin.
Then in his place at the prow of the boat, rose one of the oarsmen,
And, as a signal sound, if others like them peradventure
Sailed on these gloomy and midnight streams, blew a blast on the bugle.
Wild through the dark colonnades and corridors leafy the blast rang,
Breaking the seal of silence, and giving tongues to the forest.
Soundless above them the banners of moss just stirred to the music.
Multitudinous echoes awoke and died in the distance,
Over the watery floor, and beneath the reverberant branches;
But not a voice replied: no answer came from the darkness;
And when the echoes had ceased, like a sense of pain was the silence.
Then Evangeline slept; but the boatman rowed through the midnight,
Silent at times, then singing familiar Canadian boat songs,
Such as they sang of old on their own Acadian rivers.
And through the night were heard the mysterious sound,
Far off, indistinct, as of wave or wind in the forest,
Mixed with the whoop of the crane and the roar of the grim alligator.

THE ATCHAFALAYA LAKES, CHICOT AND GRAND OR LAKE TCHITIMACHES.

Thus ere another moon they emerged from those shades, and before them
Lay, in the golden sun, the lakes of the Atchafalaya.

Water lillies in myriads rocked on the slight undulations,
Made by the passing oars, and resplendent in heart the lotus
Lifted her golden crown above the heads of the boatman.
Faint was the air with the odorous breath of magnolia blossoms,
And with the heat of noon: and numberless sylvan islands,
Fragrant and thickly embowered with blossoming hedges of roses,
Near to whose shores they glided along, invited to slumber.
Soon by the fairest of these their weary oars were suspended,
Under the boughs of the Wachita willows, that grew by the margin,
Safely their boat was moored; and scattered about on the green sward,
Tired with their midnight toil, the weary travelers slumbered.
Swinging from its great arms, the trumpet-flower and the grape-vine
Hung their ladder of ropes aloft like the ladder of Jacob,
On whose pendulous stairs the angels ascending and descending,
Were the swift humming birds, that flitted from blossom to blossom.
Such was the vision Evangeline saw as she slumbered beneath it,
Filled was her heart with love, and the dawn of an opening heaven
Lighted her soul in sleep with the glory of regions celestial,

* * * * * * * * *

Father Felician to Evangeline:

Gabriel truly is near thee; for not far away to the southward,
On the banks of the Teche, are the towns of St. Maur and St, Martin.
Beautiful is the land with its prairies, and forests of fruit trees;
Under the feet a garden of flowers, and the bluest of heavens
Bending above, and resting its dome on the walls of the forest.
They who dwell there have named it the Eden of Louisiana.

* * * * * * * * *

Softly the evening came. The sun from the western horizon
Like a magician extended his golden wand o'er the landscape;
Twinkling vapors arose; and sky, and water and forest
Seemed all on fire at the touch, and melted and mingled together.
Hanging between two skies, a cloud with edges of silver,
Floated the boat, with its dripping oars, motionless on the water.

* * * * * * * * *

Then from the neighboring thicket, the mocking-bird, wildest of singers,

Swinging aloft on a willow spray that hung o'er the water,
Shook from his little throat such floods of delirious music
That the whole air and the woods, and the waves, seemed silent to listen.
Plaintive at first were the tones, and sad; then soaring to madness
Seemed they to follow, or guide the revel of frenzied Bacchantes;
Then single notes were heard in sorrowful, low lamentation;
Till, having gathered them all, he flung them abroad in derision,
As when, after a storm, a gust of wind through the tree tops
Shakes down the rattling rain in a crystal shower on the branches.

With such a prelude as this, and hearts that throbbed with emotion,
Slowly they entered the Teche where it flows from the green Opelousas,
And through the amber air, above the crest of the woodland,
Saw the column of smoke that arose from a neighboring dwelling:
Sounds of a horn they heard, and the distant lowing of cattle.

Near the bank of the river, o'ershadowed by oaks, from whose branches
Garlands of Spanish moss and of mystic mistletoe flaunted,
Such as the Druids cut down with golden hatchets at Gale-tide,
Stood secluded and still, the house of the herdsman. A garden
Girded it round about with a belt of luxuriant blossoms,
Filling the air with fragrance. The house itself was of timber
Hewn from the cypress tree, and carefully fitted together;
Large and low was the roof, and on slender colums supported,
Rose wreathed, vine encircled, a broad and spacious veranda,
Haunt of the humming bird and bee, extended around it.
At each end of the house, amid the flowers of the garden,
Stationed the dove-cots were, as love's perpetual symbol,
Scenes of endless wooing, and endless contensions of rivals,

* * * * * * * * *

In the rear of the house, from the garden gate ran a path-way,
Through the great groves of oaks to the skirts of the limitless prairie,
Into whose sea of flowers the sun was slowly descending.
Full in his track of light, like ships with shadowy canvas,
Hanging loose from their spars in a motionless calm in the tropics,
Stood a cluster of cotton trees, with a cordage of grape vines.

Just where the woodlands meet the flowery surf of the prairies,

Mounted upon his horse, with Spanish saddle and stirrup,
Sat a herdsman arrayed in gaiters and doublet of deerskin.
Broad and brown was the face that from under the Spanish sombrero,
Gazed on the peaceful scene, with the lordly look of its master.
Round about him were numberless herds of kine, that were grazing
Quietly in the meadows, and breathing the vapory freshness
That uprose from the river, and spread itself over the landscape.
Slowly lifting the horn that hung at his side, and expanding
Fully his broad deep chest, he blew a blast, that resounded
Wildly, and sweetly, and far, through the still damp air of the evening.
Suddenly out of the grass the long white horns of the cattle
Rose like flakes of foam on the adverse currents of ocean.
Silent a moment they gazed, then bellowing rushed o'er the prairie,
And the whole mass became a cloud, a shade in the distance.
Then, as the herdsman turned to the house, through the gate of the garden—
Saw he the forms of the priest and the maiden advancing to meet him.
Suddenly down from his horse he sprang in amazement, and forward
Rushed with extended arms and exclamations of wonder :
When they beheld his face they recognized Basil the blacksmith;
Hearty his welcome was, as he led his guests to the garden.
There in an arbor of roses, with endless question and answer,
Gave they vent to their hearts, and renewed their friendly embraces,
Laughing and weeping by turns, or sitting silent and thoughtful;
Thoughtful, for Gabriel came not; and now dark doubts and misgivings
Stole o'er the maiden's heart; and Basil, somewhat embarrassed,
Broke the silence and said, "If you came by the Atchafalaya
How have you nowhere encountered my Gabriel's boat on the bayous ?
Over Evangeline's face at the words of Basil a shade passed,
Tears came into her eyes, and she said with a tremulous accent,
Gone ? Is Gabriel gone ? and, concealing her face on his shoulder,
All her o'erburdened heart gave way, and she wept and lamented.

* * * * * * *

Then glad voices were heard, and up from the banks of the river,
Borne aloft on his comrades' arms, came Michael the fiddler.

Long under Basil's roof had he lived like a god on Olympus,
Having no other care than dispensing music to mortals,
Far renowned was he for his silver locks and his fiddle,
"Long live Michael," they cried, "our brave Acadian minstrel!"
As they bore him aloft in triumphal procession, and straightway
Father Felician advanced with Evangeline, greeting the old man
Kindly and oft, and recalling the past, while Basil, enraptured,
Hailed with hilarious joy his old companions and gossips,
Laughing loud and long, and embracing mothers and daughters,
Much they marvelled to see the wealth of the ci-devant blacksmith,
All his domains and his herds, and his patriarchal demeanor;
Much they marvelled to hear his tales of the soil and the climate,
And of the prairies, whose numberless herds were his who would take them;
Each one thought in his heart that he. too, would go and do likewise.

* * * * * * *

All was silent without, and, illumining the landscape with silver,
Fair rose the dewy moon and the myriad stars; but within doors,
Brighter than these, shone the faces of friends in the glimmering lamp-light—
Then from his station aloft, at the head of the table, the herdsman
Poured forth his heart and his wine together in endless profusion.
Lighting his pipe that was filled with sweet Natchitoches tobacco,
Thus he spake to his guests, who listened, and smiled as they listened:
Welcome, once more, my friends, who so long have been friendless and homeless,
Welcome once more to a home that is better perchance than the old one.
Here no hungry winter congeals our blood like the rivers;
Here no stony ground provokes the wrath of the farmer,
Smoothly the ploughshare runs through the soil like a keel through the water—
All the year round the orange groves are in blossom, and grass grows
More in a single night than a whole Canadian summer.
Here, too, numberless herds run wild and unclaimed in the prairies.

CALCASIEU PARISH.

LAKE CHARLES

The town of Lake Charles is situated on the northeast side of the lake of the same name. It has a handsome site, and commands a full view of the lake, and of the inlet and outlet of Calcasieu river. The town contained 502 inhabitants in 1875. It contains twelve stores, two bakeries, 2 barber shops, two blacksmith shops, 4 bar-rooms, two butcheries, a shipyard, a tanyard, a taylor, a shoemaker, a tinsmith, three lawyers, two physicians, a drug store, a photographic gallery, a dentist, a teacher of instrumental music, a planing machine, a steam saw-mill, two painters, a brick-mason, two carpenter shops, a post-office, a telegraphic office, a printing office (The Echo,) three schools, one Catholic church, one Methodist church, one German church, a Masonic hall. etc., etc.

THE LAKE.

The Lake is about circular in shape, and nearly two miles in diameter. The water is always fresh, and generally clear. The banks are alternately high and low. In some places the banks are twenty feet above the level of the Lake.

HEALTH.

The town of Lake Charles is said to be exceedingly healthy, and a pleasant summer residence; the sea breeze, and the breezes from the open prairie fan the town at all seasons of the year. The thermometer seldom or never gets above 96 degrees, seldom above 90, and it is not often down to the freeziag point.

RAILROADS, SAW-MILLS, ETC.

The projected New Orleans and Texas Railroad passes through the town.

Within three miles of the town there are nine steam saw-mills and five or six stores. The country in the vicinity of Lake Charles is pretty thickly settled, considering its remoteness from the great thoroughfares.

The numerous saw-mills of Calcasieu river give steady employment to a fleet of fifty or sixty schooners, with a capacity each of from ten to a hundred thousand feet of lumber. This lumber is carried to the ports of Texas and of Mexico.

Four or five steamers ply on the Calcasieu river, and all are engaged in the lumber business but one which carries the U. S. mail twice a week from Lake Charles to Cameron, and to Calcasieu Pass, a distance of fifty miles.

The principal business of the parish is the lumber interest, next cattle raising, and last farming. The land is usually poor, and all of the fertilizers applied is by cow-penning.

MINERALS.

Gypsum, sulphur and petroleum are found in Calcasieu, not far from Lake Charles, and near the west fork of the Calcasieu river. It has been well known for many years that petroleum existed in this parish. An oil company was formed since the war to search for oil near where it has appeared at the surface for many years. In boring they struck sulphur, and further down, gypsum.

The borings show:

	feet.
Soil	2
Solid clay, intercepted with 2 strata of quicksand, 22 and 15 feet thick	163
Quicksand	179
Crumbling marl,	2½
Calcareous sand	30½
Calcareous marl, with pebbles	4
Hard, compact, calcareous stone	5
Pure, white, saccharoid, caleareous substance	42
Sulphur (77 per cent. pure sulphur)	112
	540
Gypsum	700
	1240

RICE CULTURE.

There is much land in the parish that is too low and wet for corn, cotton, or sugar cane, that would make splendid crops of rice. Attention is being turned to this branch of farming.

ORANGE GROVES.

Two of these orange groves produce yearly about 100,000 oranges each. The yearly crop is about 350,000 oranges, or about 1000 barrels.

In the last five or six years,. a great impetus has been given to orange culture in this parish. The trees are healthy and have not been troubled with the diseases, and enemies which have preyed upon the Florida orange groves. Some of the orange trees of this parish have trunks a foot in diameter, and an altitude of 25 feet. Some trees are said to bear from three to five thousand oranges to a tree yearly. The fruit sells at from $12 to $15 a thousand, with the demand constantly increasing. It is safe to predict that in four years the orange groves or Lake Charles will produce more than a million of oranges yearly. The fruit is all shipped to Texas, which is now the only market.

DISTANCES AND FARE.

	Miles	Fare
New Orleans to Brashear, (R. R.)	80	$ 4
Brashear to New Iberia (Steamboat)	70	4
New Iberia to Lake Charles (Stages)	115	12
	265	$20

Time, 50 hours.

In lumber schooners, between Lake Charles and Galveston.

	Miles
Galveston to mouth of Calcasieu,	90
Mouth of C. River to Lake Charles,	50
	140

Fare from $5 to $10. Time 24 hours, or more, according to wind; sometimes several days.

Distance between New Orleans and Lake Charles by the projected line of railroad 210 miles.

CAMERON PARISH.

GENERAL FEATURES.

Cameron is the most southwesterly parish in the State—bounded west by Sabine Lake and river, north by Calcasieu, east by Vermilion, and south by the Gulf of Mexico. It is mainly composed of two belts of land, extending with occasional interruptions, nearly the full length of the parish from east to west. The larger portion of the parish is sea marsh; much of it is almost impassable, and as yet unexplored. This marsh has a width of about twelve miles.

DAIRYING.

Cameron parish is said to be admirably adapted to dairying. The grass is abundant and nutritious the year round. The climate is exceedingly mild in winter. The sea breezes and the great evaporation from the large bodies of water in and around the parish moderate the heat of summer. Some of the Creole cows give twelve quarts of milk a day without shelter in winter or summer, and with no food except the natural grasses of the marsh and the prairies. As good butter has been and can be made along this coast, as is made in the North or West.

CATTLE.

It is impossible to approximate a correct estimate of the quantity and value of live stock annually pastured and fattened for market within the limits of Cameron parish. For stock raisers these marshes are an exhaustless mine of wealth. Their luxuriance as pasture surpasses any powers of description at command of an ordinary writer. The wild pea vine mats over thousands of acres of these marshes, and the cattle revel in these vines, and enjoy the feast exceedingly. The variety and richness of grasses and herbage for cattle found in these sea shore pastures are surprising.

FLIES AND MOSQUITOES.

The flies and mosquitoes are troublesome, but the cattle instinctively herd together in seasons when flies are most harassing, and by chafing against one another, and by their breath, which these pests dislike, they secure considerable protection against their attacks.

BAD STOCK WATER.

The want of good water in some ranges is a severe drawback in stock raising, but this will doubtless be in a great measure obviated by the use of wells and windmill pumps, and tanks, as the country becomes settled up with men of enterprise, and as a more advanced state of farming and stock raising is reached. Wholesome fresh water can be had in these coast ridges in from six to ten feet below the surface. Even on some of the sandy islands along the coast wells of tolerably good fresh water are obtained by digging from five to ten feet, with salt tide-water not a hundred feet distant.

We believe that domestic animals along this coast, when properly cared for and attended to, are exceedingly healthy, and that the diseases and mortality among them are usually attributable to neglect and want of proper attention,

SHEEP.

Sheep are said to be profitable in Cameron, though flat lands are generally not favorable to sheep. The sheep here are generally free from disease, and are not troubled with ticks, or destroyed by wolves. The sheep are small, no attention having been given to improving the breed; but the wool usually brings about thirty cents a pound. Mutton, in pretty large quantities, is yearly shipped from Cameron, and brings in a handsome revenue. Sheep raising is here more remunerative than cattle, or any other stock Farmers are yearly increasing their stocks of sheep by purchase, as well as natural increase.

HOGS.

The marshes of Cameron are the paradise of hogs. Thousands of pounds of pork are secured yearly by simply marking the ears of the pigs, and shooting the hogs when fat. The marsh, and its bayous and waters, are full of various kinds of fish, and food for hogs;—crawfish, fiddlers, crabs, clams, shrimps, and vermin of numerous species. In the spring the marshes abound with birds of various kinds, and eggs. The ground is filled with roots, and covered with grass and vegetables which hogs are fond of. The hog range is immense, and most excellent. Hogs do not mind mosquitoes, and they devour the marsh snakes when they come in their way. And such splendid wallowing in the marsh mud! Hogs would never willingly emigrate from such a land of promise. And they find abundance of food where horned cattle and sheep would sink in the mire.

POULTRY AND EGGS.

Poultry succeed well in Cameron, and all along the Gulf coast. A single schooner sometimes takes off a dozen barrels of eggs at once to the Galveston, or other markets, and coops of chickens besides; very good shipments, considering the sparseness of the population.

WILD FOWLS.

Water fowls are abundant, in winter, in the lakes and bayous of Cameron. The range is as favorable to wild ducks, geese, and brant, as to hogs, deer, otters and muskrats.

HOME FERTILIZERS.

Cameron parish is prolific in home fertilizers. Marsh muck is excellent when composted with lime and manure. The supply is exhaustless. There are large shell banks from which fertilizers may be drawn, and the wild and domestic pea vine, cow pen manure, bones, etc., may be used.

"BACK LANDS."

The great body of arable lands in Cameron are what are called "back lands." They are tenacious, full of tough grass roots, and it is hard work for two yoke of oxen to break them up. But, when subdued and exposed to the sun and weather, and the roots decayed, they are very productive. They can only be ruined by bad cultivation and abuse, while, by judicious farming, they would retain their richness for several generations with but slight expense for fertilizers.

BAD TEAMS.

Ox teams are used in this parish for farm and other work. The oxen are not large, and are badly broken. It is not uncommon to see three yokes of oxen in a team, with a rope to each of the nigh ox's horns.

The horses are usually the small creole breed.

SUGAR CANE.

Sugar cane grows finely in Cameron, but there is a scarcity of wood. In the pine lands of Calcasieu, near by, they have plenty of wood, and by cow penning their land, or putting on it suitable fertilizers, they can make two hogsheads of sugar, and three barrels of molasses, to the acre.

POTATOES.

Sweet and Irish potatoes do finely in Cameron. Irish potatoes are planted largely for sale, and have at times been the most remunerative of all crops in proportion to the area. The cost to the farmer is almost nothing. Irish potatoes are frequently planted in December, and are ripe in May, in season to plant, on the same ground, corn, cotton, or sweet potatoes, or highland rice, and some other things. The yield, from even careless planting and cultivation, is from twenty to sixty barrels to the acre.

The Honduras, six weeks, or hog yam, has been a great benefit to the farmers of Cameron. They are not very good for table use, but are excellent for stock and hogs. They grow to a mammoth size in a short time. It is supposed that they will yield eight hundred bushels, or more, to the acre.

ONIONS.

Onions yield bountifully in Cameron, and it is thought they may be made a valuable stable production of the parish for other markets.

MELONS.

The melon crop of this parish is usually fine. No better melons can be raised in any part of the State, or any where along the coast, east or west.

GARDEN VEGETABLES.

Garden vegetables grow to great perfection in this parish. Any vegetables that succeed anywhere in the south do as well here as in any other locality.

FRUITS IN CAMERON PARISH.

ORANGES.

Oranges are the fruit of fruits in Cameron. Though other fruits do well, in comparison with this queen of fruits all others sink into insignificance. It is claimed that an individual tree in Cameron, in a single year, produced *eight thousand merchantable oranges*, which brought in Galveston one hundred and fifty dollars. We do not vouch for the truthfulness of this statement,—we take it from an article published in the WEEKLY ECHO, at Lake Charles, on the 8th of July last, dated at Leesburg, Cameron parish, we trust the authority is reliable and the statement true. The orange crop of Cameron is estimated to be worth $10,000 annually, and in a few years, trees planted on and east of the Mermentau river, will return to that section $30,000 annually. Orange culture is here in its infancy.

VARIOUS FRUITS IN CAMERON.

Plums are to be had from the first of May until after peaches are ripe, in abundance,

Grapes grow wild in profusion everywhere that they can get a foothold. The cultivated varieties have received but little attention. They do well when properly cared for.

Blackberries and dewberries grow wild in abundance, but the cultivated varieties have never been planted here.

Strawberries grow in great perfection.

Peaches have received little or no attention here. Those trees that have been attended to as they should be thrive well. Most of the peach trees are seedlings, and from inferior fruit. Delicious fruit from grafted trees are produced in the parish, and may be produced in abundance.

But little attention has been paid to the cultivation of any fruits in the parish except oranges, and these only for a few years.

LOUISIANA PLANTATIONS AND FARMS.

Louisiana had, in 1870, plantations and farms as follows:

Over 1000 acres each,	142
Between 500 and 1000 acres,	650
Between 100 and 500 acres,	3,753
Under 100 acres,	23,936
	28,481

Lands in cultivation in 1870, 2,045,670 acres. Tillable lands not cultivated, 18,000,000 acres.

SUGAR PLANTATIONS, CROPS, ETC.

E. J. Forstall, Esq., in DeBow's Review, September, '66, makes the following estimates:

Number of sugar estates in operation in Louisiana in 1861 and '62, 1292; number in operation in 1863 and '64, 180; capital invested in sugar interests in Louisiana in 1861 and '62, $200,000,000. Slaves fed and clothed by sugar planters, 139,000.

From the Richmond Enquirer:

That the rich lowlands of Louisiana are capable of producing all the sugar required for consumption in this country is beyond question. As yet their capacity in that direction has been tested only in the slightest degree. The high standard of Louisiana sugar and molasses is universally admitted, and if the productive resources of the delta were brought into complete requisition there would be little necessity of obtaining sugars from abroad. If the war in Cuba be continued a few years longer her sugar crop must be greatly reduced. In view of the contingencies which may then arise it would be well for the far-seeing capitalists of the West and North to invest in Louisiana sugar lands, now that they are selling for a song because of the long continuance of Radical rule, a rule now nearly at its close.

CHEAP SUGAR ESTATE.

SHRINKAGE.

NEW ORLEANS, September 6, 1872.

To the Editors of the Picayune:

In the Times of to-day there is an article comparing city taxes with those levied in the country parishes, wherein ground is taken that assessments in the country are not only comparatively but are absolutely too low. Facts are at variance with the figures given, as there is no doubt that the depreciation of country property is greatly in excess of the difference in the assessments of the years 1860 and 1871, upon which the dictum of the Times is based. The assessments for the above mentioned years were respectively in round numbers, $470,000,000 and $250,000,000.

As sugar plantations and negroes constituted much the greater part of the taxable property in the rural districts prior to the war, a few facts in relation to the present and past value of sugar estates will exhibit that if the assessments of 1860 were even approximately correct, those of 1871 are ruinously excessive.

On the death of Mr. Henry Doyal, a well known planter in Ascension, his property, consisting of four or five sugar estates, was valued at $1,800,000, but was known to be worth in the market $2,500,000. This property would not now sell for $100,000, basing my opinion upon the value of two of these plantations, recently sold.

The model plantation of this State for many years was that of the late Mr Valcour Aimé, of St. James. With its substantial and elegant improvements, this plantation could not have been bought before the war for $1,000,000. For the purpose of settling the succession of the estate, it was offered for sale twice last winter, but no bidder was found to offer the value to which it was limited--$75,000, I believe. It was afterwards sold for $50,000.

Linwood plantation, in Ascension, not in cultivation since the war, worth at least $200,000 in 1860, was sworn to, by one of the most sagacious planters and merchants in the State, as not being worth $10,000. In the opinion of the witness, it could not be sold at any figure for cash.

These few instances are a fair indication of the depreciation of the value of property in the country. Plantations will—except in occasional instances—sell for a little more than the value of the movable property on them.

WHY THESE LANDS ARE CHEAP.

Strangers may ask why these lands have depreciated so much in their market value. It is because forced labor has been suddenly abolished; and we are going through a transition that involves sacrifices and ruin; because we have had ruinous taxation, and our people have not been prompt to meet the new order of things and work unitedly, and with power for the public good. There are farmers in the sugar, and rice and cotton regions of this State, who have made money in spite of all these discouragements, and who would not now sell their lands for a hundred dollars an acre; whose lands to them are worth a hundred dollars an acre, and more. All of these sugar and rice lands would be worth a hundred dollars an acre if they were all cultivated by farmers like scores that we are acquainted with and could name.

VAST AGRICULTURAL RESOURCES—MR. DARBY'S VIEWS.

Darby, a native of Pennsylvania, surveyed all of Louisiana which lies between the Mississippi River and the Sabine, over fifty years ago. He knew more about the real merits of the State, so far as its soil was concerned, than any living man. His labors in surveying the State necessarily had to be done on foot, and he camped on nearly every lake, bayou, river, prairie, and hill, and in all the swamps and flat lands of the State. In his geographical description of Louisiana, published soon after he made his survey, he remarks:

"When the State of Louisiana is brought into cultivation far from the maximum, four hundred thousand laborers may be employed in making cotton, three hundred thousand in the cultivation of sugar cane, and one hundred thousand in the production of rice. This working population of eight hundred thousand people, at the most reduced prices of the different products, would yield one hundred and thirty million dollars. When the elements are examined separately, this rich reward that is offered to the industry of man by the soil of Louisiana, will not appear delusive."

THE FACTS BETTER THAN THE PREDICTION.

That Darby greatly underrated the capacity of Louisiana soil, the following table, showing the sugar crop of 1861, and other crops of 1860, will clearly prove.

These crops are on record, as follows:

459,410 hogsheads of sugar.
700,000 barrels of molasses.
777,738 bales of cotton.
6,331,257 pounds of rice.
39,940 pounds of tobacco.
16,853,745 bushels corn.
2,060,981 " sweet potatoes.
294,065 " Irish potatoes.
431,148 " peas and beans.
89,000 " oats.
1,444,742 pounds of butter.

Products of gardens and market gardens $527,000
Value of home mnnufactures 502,000
Value of animals slaughtered.............................2,095,000
The land in cultivation in Louisiana when the above crops were made did not exceed (acres) 2,700,000
The entire area of the State, not including lakes, rivers, bayous and bays, is (acres)..........................26,105,600

VALUE OF YIELD PER ACRE AND PER HAND.

The above products, and all other agricultural productions of the State in 1860, at the usual market prices, would bring about $125,000,000. The five staple articles, at eight cents a pound for sugar, fifty cents a gallon for molasses, $50 a bale for cotton, and the usual price for rice and tobacco, would bring nearly $93,000,000.

WORKING FIELD FORCE.

It would be a large estimate to put the working field force of Louisiana before the war, including all colors, at $125,000 hands, out of a rural population of about 500,000 souls.

$1,000 PER HAND, AND $46 PER ACRE.

Estimating the crop of all kinds, and all agricultural productions in a single year to be worth $125,000,000, this would give $1,000 a hand to every field hand in the State, and $46 an acre to every acre in cultivation. The staple crops alone, sugar, cotton, etc., worth $93,000,000, would give, divided equally among all of the field laborers of the State, $744 to the hand; though many hands did not cultivate the staple articles.

VALUE OF FARMS AND CROPS COMPARED.

In 1870, the farms, farming utensils and machinery of Louisiana were valued at $75,374,000. Crops, same year, $52,000,000.

Value of Louisiana crops, as above, compared to value of farms, etc.		69	per cent.
In Iowa, same year		28	"
" Illinois,	"	22	"
" Ohio,	"	19	"
" Indiana,	"	18½	"

And Louisiana, at that time, five years after the war, had the disadvantage of a new and unsettled system of labor, had bad teams, and bad utensils, but little money, no credit abroad, a bad state of government, high taxes, a multitude of the most crushing discouragements to contend with, while the Western farmers had none of these evils before them.

SOIL RESOURCES.

It may be interesting to strangers who do not appreciate the immense and varied resources of Louisiana to examine a brief, though imperfect, catalogue of the articles which industry, skill, enterprise and capital may secure in abundance in this State, for the benefit of our own people, or as articles of commerce.

A model Louisiana farm may furnish beef, pork, mutton, veal, butter, cheese, eggs, chickens, etc., in abundance. Also corn, wheat, buckwheat, oats, rye, barley, millet, hay, beans, peas, broom-corn, castor oil beans, cotton, sugar, rice, tobacco, pumpkins, sweet and Irish potatoes, pindars, chofas, melons, red and other clovers, grasses of numerous kinds; ramie, jute, flax, cashaws, gourds, bene plant, indigo, arrow root, ginger, cabbages, turnips, okra, Jerusalem and bur artichokes, cucumbers, onions, and all other garden vegetables that can be produced in the Northern States.

All the fruits of the Northern States, except perhaps gooseberries, damsons, and one or two others, can be raised in the northern part of this State, and many tropical fruits in the southern part, as we have elsewhere noticed more particularly.

THE BILL OF FARE OF A MODEL LOUISIANA FARMER.

A model farmer in Louisiana may have an abundance of the raw materials named above, and should make his year's supply of grape and blackberry wine, and other kinds of wine; also cider, many kinds of preserves and jellies, and syrups of different flavors, all made of home produced fruits and sugar; also honey of excellent quality, and plenty of delicious butter the year round; he should have home-made, home-dried and home-canned fruits of various kinds in abundance the year round, and home-made peach and apple brandy, and home-made brandy fruit, (if distilled spirits are used on the place.) He should have wool and cotton to send to Southern factories in exchange for Southern made cotton goods and cloth. A model Louisiana farmer should produce more of the comforts and luxuries of life than the best farmer in any other State in the Union.

SILK WORMS IN LOUISIANA—TRADE IN SILK WORMS' EGGS—LOUISIANA COCOONS—SUMAC—SUNFLOWER—CASTOR OIL BEANS.

SILK WORMS.

Silk worms feed on mulberry leaves twenty-five to thirty days and then make cocoons in twenty-four hours. They eat out of the cocoon in eight days. Each female lays about 400 eggs. To use the cocoon for silk the worm must be destroyed by heat before he eats out. They hatch when the mulberry leaves appear in spring.

After they leave the cocoon, in the winged or transformed state, they at once lay their eggs, and both males and females die.

One hundred females lay an ounce of eggs, worth in the market more than their weight in gold. After they hatch in spring, in sixty days they make cocoons.

LOUISIANA COCOONS AND SILK.

An Italian who has made considerable money in producing silk worms' eggs in Louisiana for the Italian market, informed us that
4½ pounds Louisiana cocoons make 1 pound of silk.
12 to 15 pounds of Japan cocoons make 1 pound of silk.

Women and children can do all the work in feeding silk worms and in reeling silk from the cocoons.

They can make in Louisiana at this business $100 to the hand in two months, in the silk-worm season.

Raw silk, reeled from cocoons, finds a ready market and brings good prices in New York.

Silk culture was a business of some importance among the early French settlers of Louisiana.

TRADE IN SILK WORMS' EGGS.

Europe pays Japan four million dollars yearly for silk-worm's eggs. Louisiana can produce these eggs in immense quantities. The Italian above referred to sends the eggs to New York. If he wishes to make silk he has the amount of eggs he desires sent back by express, when the mulberry trees put on their spring robes.

Thiers estimated the loss to France by the silk-worm disease at twenty million dollars yearly. We have no such disease in Louisiana, and the silk worm of this country is larger and more valuable than the silk-worm of France and Italy.

SILK AND COTTON.

In eighty years silk culture in France advanced from five to one hundred and fifty million dollars, and the advance in cotton culture in the United States was about the same.

In 1792 the question was seriously discussed in England, whether the United States would ever make one hundred bales of cotton. 5,000,000 bales have since been made.

SUMAC.

As oak and hemlock bark in the Northern States and Canada, their tannin material, is nearly exhausted, the sumac question and tannin materials in the Southern States will soon become highly important.

Upland sumac is that preferred for market purposes. It is of three species: the stag horn, or *rhus typhina*, which grows to a tree eighteen or twenty feet high, and the smooth *rhus glabra*, the *rhus capallina*. This is sometimes called dwarf sumac. The swamp sumac is not considered valuable.

The first has branches of a hairy appearance, like a stag's horn; the two latter have little bristles on the berries. These sumacs generally grow on red lands derived from primitive rocks. The leaves only are used. They are dried, ground, and put in bags, and sell in the New York market at 4½ to 5 cents a pound, if carefully saved and prepared.

Southern sumac is richer in tannin than the Sicilian, but there is more money in hay, ground peas, or most any of the Southern farm crops than in sumac at present prices. It may pay well in a few years.

THE SUNFLOWER.

The sunflower grows finely in Louisiana. It yields fifty bushels of seed to the acre. The seeds yield a gallon of oil to the bushel, said to be equal to almond or olive oil. Fowls are fond of the seeds, and they are as valuable for many purposes as corn. From 15,000 to 20,000 plants can be grown on an acre.

CASTOR OIL BEANS.

In St. Louis, in 1867, about 50,000 bushels of castor oil beans were sold at $1 25 to $3 90 a bushel. They yield sixty gallons of oil to the acre, worth $1 50 a gallon. Castor oil beans grow as well as any other crop in Louisiana, and may be made profitable in connection with other crops.

TEA SHRUB IN LOUISIANA.

Mr. Manard has succeeded well in cultivating tea in Tangipahoa parish, on the New Orleans, St. Louis and Chicago Railroad, in this State. He has made over twenty-five pounds in one season, and could increase the crop to any desired extent. We have enjoyed as good a cup of home-made tea at Mr. Manard's table as we would wish to enjoy at any man's table. We could not distinguish between the merits of his tea and the best teas from the East. He prepares the leaves with great care, works out all of the gummy ingredients, and is well informed in regard to the East India process of curing tea in the best possible manner. The crude leaves, without the workings and manipulations which he carries them through, make no better tea than sage or mint.

ELDERBERRY WINE.

Elderberry wine is said to be highly prized by the English. There are elder orchards in New Jersey, where they make this wine, and they find a ready sale at remunerative prices for all they make. Where does the elder and elderberry grow more luxuriantly, in defiance of weeds, bushes, deluge and drouth, than in Louisiana?

MOSS AND PAPER STOCK.

Southern moss has become a valuable article of commerce. It

hangs upon the trees in our swamps and low lands in millions of tons. Instead of producing malaria, as some suppose, it doubtless absorbs and feeds upon malaria and renders the air more pure and healthful.

The supply of paper stock in this State is almost limitless. Palmetto leaves and stalks make good paper. Bagasse or pressed sugar cane, makes good paper. Mixed with cotton rags, it makes an excellent newspaper material. Wild cane has been manufactured into excellent paper. Cotton stalks and okra stalks, bear grass and many other articles that may be furnished at cheap rates in Louisiana make excellent paper.

CONCLUSION.

What State or what country can furnish so lengthy, rich and varied lists of agricultural articles of value as Louisiana? With a rich soil and unlimited supply of natural fertilizers, the best in the world, what prevents our State from being rich? It is because—

1. Not a seventh of our land is cultivated.

2. Out of a population of over seven hundred thousand, not one-seventh of our people follow the plow and work the hoe.

3. The most valuable resources of the State have hardly been touched; the small industries of the State, which must lie at the foundation of its future prosperity and wealth, have been treated as though they were unworthy the notice of the capitalist or the legislator.

4. When the soil of Louisiana is cultivated by a million, in place of a hundred thousand ploughmen; when her fertilizers are judiciously applied; when her fruit, and grass, and stock, and farm crop capacities are properly tested by the best class of modern farmers; when our mines and mill sites are all occupied and worked; when New Orleans wakes from her slumbers and opens the passage to the sea, and estabishes the barge system on her twenty thousand miles of navigable waters, between the Alleghany and Rocky Mountains, then the world will have more correct views than the present generation has of the priceless natural wealth of the forests, and soil, and mines of the State, and the Crescent City will have to reclaim the swamp lands around her, to make room for her million inhabitants—or, so it appears to the writer.

THE RICH LANDS OF LOUISIANA.

WEALTH IN THE SOIL.

More than twenty million acres of lands of this State are naturally rich, though many million acres of these rich lands are not arable.

	ACRES.
Arable alluvial lands of the State,	3,615,000
Wooded alluvial or swamp lands,	2,752,000
	6,367,000
The prairies of Attakapas, and of St. Landry and Sabine parishes, most of them rich	2,483,000
Coast marsh, all rich,	3,338,000
	12,188,000

Prof. Hilgard, in his "Preliminary Report of a Geological Reconnoissance of Louisiana, in 1869," speaking of the coast marsh, says:

"The soil of these lands, being of great depth, and composed of rich organic mold, whenever it has been drained and brought under the plow, has surpassed all other land in the State in fertility and productiveness. It is in such lands that the culture of rice has attained so much success. Lying, as this region does, in the extreme southern limits of the State, on the warm waters of the Gulf of Mexico, and protected on the north by the numerous broad lakes which overhang the coast, the climate is semi-tropical, and admits of the culture of frui;s and productions that are strictly tropical."

"No field," says Prof. Hilgard, "can offer a more promising prospect to capitalists, agricultural companies and immigrant aid societies, than the cheap and easy reclamation of these rich marshes. The market price of the lands is twenty-five cents an acre, and they may be made equal to any in the State by the use of modern steam dredging machines for canaling and leveeing them, and the improved wind mill pumps for draining them; for the breeze is constant near the seacoast.

"Meteorological records kept below New Orleans for seventeen years, show that we do not average five days in the year without wind;

and that usually there is a stiff breeze, lasting the greater part of the day, which greatly moderates the heat and tempers of the cold."

The good uplands of the State comprise, ...5,248,000
The bluff lands, ...1,585,000

6,823,000

These are generally excellent farming lands and produce fine crops in the hands of good farmers. And there are large tracts of excellent land, even in the pine regions. Most of the pine lands have retentive and superior subsoil, and where green crops are plowed in and manures and home-made fertilizers used, as they are applied on the old lands in the older States, they are capable of producing a bale of cotton, or thirty-five bushels of corn to the acre, and other crops in proportion.

IMMENSE AND EXHAUSTLESS NATURAL WEALTH.

LOUISIANA.

The immense natural wealth of this State consists in —

1. Its vast forests of excellent timber, and the great value of many of these varieties of timber.

2. Its rich and diversified soil, well adapted to almost all of the agricultural productions, the field and garden crops and grasses, of colder climates, and most of the fruits of higher and lower latitudes.

3. Its salubrious climate, congenial to both man and to domestic animals and fowls.

4. Its 275 bayous, 115 lakes, 56 bays, and 40 rivers, and water surface of 2,300,000 acres, furnishing extensive and convenient navigation, and numerous mill sites. The waters of Louisiana contain great values in their fishes, turtles and oysters; besides a vast value in moderating and mellowing our climate in winter, and by evaporation, moderating the force of the summer sun and extracting malaria from our atmosphere.

Oysters, fish, and water game, may be furnished in abundance in their proper season for our local markets; fine sea turtle may be taken

9

in large numbers on the islands that line the Louisiana coast, and these and shrimps and oysters may be canned on an extensive scale for the markets of the world.

The principal fishes in Louisiana waters are the red fish, black drum, flounder, sheephead, trout, croaker, mullet, buffalo, perch, gar, choupique, casbugo, sacalait, bar, pike, white cat,; also soft-shell turtle, shrimps, crabs, etc.

5. Its vast mine of pure salt, its salt springs, its sulphur mine, its undeveloped wealth in coal oil; its coal and lignite, iron, lime, soda, gypsum, and potter's clay; and fine materials for building, paving, and fire bricks.

6. Its local advantages at the mouth of the finest navigable river on the globe, which has 20,000 miles of navigable waters.

WEALTH IN THE FORESTS.

FORESTS, TIMBER, SHRUBS, ETC.

Louisiana has, in her forests, values which her citizens but poorly appreciate. While the Northern and Western States have nearly exhausted their lumber resources, their oaks and hemlocks, which were their chief reliance for tanning materials, Louisiana has immense forests of the finest timber on this continent, including numerous varieties of oak. We give the lists below:

OAKS, eighteen varieties, as follows: Live, white, red, black, brown, scarlet, water, willow, overcup, Spanish, Dentata, chincapin, swamp, bear, turkey, myrtle, black jack and post oak.

CYPRESS, two kinds: red and white.

PINE, three kinds: long leaf, short leaf and loblolly.

ASH, four kinds: réd, white, water and green.

ELM, three kinds: water red and slippery.

GUM, three kinds: black, sweet and tupelo.

HICKORY, four kinds: shell-bark, black, pig-nut, and water.

MAPLE, three kinds: sugar, silver, and swamp.

LOCUST, three kinds: honey, black and acacia.

MULBERRY, two kinds: white and red.

SASSAFRAS, two kinds: white and red.

MAGNOLIA, two kinds: grandiflora and glauca.

PECAN, six varieties.

CHINA, three varieties.

WILLOW, four kinds: red, white, black and weeping.

Also, black walnut, cotton-wood, yellow popular, beech, osage, orange or bois d'arc, sycamore, hackberry, catalpa and linn or bass wood.

The following is a list of the smaller trees and also shrubs.

Wild peach, or mock orange, balm of Gilead, prickly ash, (two varieties,) sumach ash, (two varieties), wild cherry, wild plum, dog wood, (two kinds), holly, box-elder, box-wood; red, may, apple, sugar and parsley-leafed, bird and black haw; red bay, sweet bay, crab apple, (two varieties), persimmon, (two varieties), iron wood, yellow-wood, wax-myrtle, horn-beam, buckeye, Yaupon, sorrel tree, wild sloe, hasberry, holly, red bud, water birch, buckthorn, papaw, alder, (two kinds).

The above list is from "Darby's Geographical Description of Louisiana," with some few additions made by the writer, and there is an unwritten list of the smaller growth, shrubs, vines, etc., yet to be added by future writers upon this subject.

THE MOST VALUABLE TIMBER.

Cypress trees in millions of acres, the finest building timber in the South, the best for frames, floors, weather-boarding, doors, sash, blinds, fences, and other useful purposes, stand upon Louisiana soil, and usually near good navigation.

Oaks, valuable in ship building and for staves as an article of commerce, for manufacturing into carts, plow beams and handles, wagons, utensils and machines of various kinds, for cabinet purposes, and for thousands of useful purposes in all civilized and enlightened nations; its bark of great value for tanning purposes. Oaks of numerous varieties and in vast numbers are growing in the forests of this State.

Hickory trees are found in this State in vast numbers, and of fine size and quality. Beech, far taller and larger than those of the Northern States, are found in great numbers in the bottom lands of this State. It is a valuable timber.

Large quantities of ash, often of surprising growth, grow in the swamps of the State. The great value of this timber is familiar to all.

Several kinds of gum, in vast quantities, are found in Louisiana.

When seasoned, it is light and tough, and makes superior wooden ware, bowls, trays, and other articles. The root of the tupelo gum, which grows in our wet swamps, is a fine substitute for cork. It is as light as cork when first taken from the live tree, and mud, and water. Magnolia has a white, pliable grain, and the shavings from the edge of a plank have been worked into neat baskets of various kinds.

The pine found in Louisiana in millions of acres, sometimes four feet in diameter and a hundred feet high, is valuable for building, fencing and many other purposes.

Our pine forests may furnish large quantities of pitch, tar, turpentine, rosin and charcoal.

Pine lumber of excellent quality is sold at the mills of this State at $10 a thousand feet, and cypress often sells for $15.

Millions of feet of pine and other kinds of valuable timber are yearly destroyed by fire in clearing lands, while in the West wood is so scarce that they sometimes burn corn for fuel, and their far-seeing men are alarmed at the lumber famine that now threatens them.

Cherry, walnut, cedar, china, poplar, pecan, elm, catalpa, all are valuable timber, and are found in abundance on this prolific and valuable soil.

There are many kinds of timber in the forests of Louisiana that may be valuable for cabinet work.

From the wax myrtle berry large supplies of tallow may be obtained of a quality, for some purposes, superior to sperm.

Many valuable qualities could be mentioned in relation to other trees and numerous shrubs that are found in the forests, swamps, marshes and praries of this State.

OAK STAVES.

From the New Orleans Times.

The oak stave business is by no means an unimportant consideration. Messrs. Robet Brothers informs us of the following facts, showing the receipts of staves during the year:

Total pipe staves	2,335,622
Total short staves	1,096,106
Total	3,431,728

From Mr. J. H. Mehaffey we have the following statement of the staves sold to the city trade, on hand and for sale :

Sold to the city trade—985,899 pipes, 68,309 hogsheads, 118,529 clarets, 14,657 barrels. On hand and for sale—302,271 pipes, 107,585 hogsheads, 28,993 clarets, 8281 barrels, 1023 kegs, 2520 headings. Total pieces, 1,938,037.

The average value of these staves is $100 per thousand.

AREA OF LOUISIANA FORESTS.

The value of the forests of this State can hardly be over-estimated. This State contains, independent of bay and inland water surface, an an area of

		Acres.
		26,105,600
Our prairies cover	2,483,000	
The sea marsh	3,338,000	
All cultivated lands in the State before the war, including the prairie lands	2,700,000—	8,521,000
		17,479,000

WOODLAND.

Towering and noble pines cover six millions of acres, but the pines are often mixed with oak and other valuable timber. The State contains 2,752,000 acres of wooded alluvial, or swamp lands, covered with cypress trees of fine size, mixed with cottonwood, ash, oak, sweet, black and tupelo gum, honey locust, slippery elm, swamp, white, Spanish and water oak, and live oak, pecan, elm and other kinds of forest trees. There are 5,248,000 acres of good uplands, for the most part covered with oaks of various kinds, mostly red, white, black and post oak; also hickory, black and sweet gum, maple, ash, sassafras, beech and dogwood, and some smaller trees; also prickly ash and hackberry, and fox grapes and muscadines. As not an eighth of Louisiana has ever been in cultivation, most of the lands, except the sea marsh and prairie, are covered with timber.

WOODLANDS AND FUEL.

While fuel is becoming scarce all over the Northern States, and has always been scarce in the Western prairies, a farmer on the Attakapas prairies, and in other portions of the State, may plant china berries with his corn, get two successive crops of corn, and the third year, ce ten acres of china trees, 100 trees to the acre, he can get his supply of firewood from the trimmings of his trees, and an ample yearly supply during a lifetime; and by seeding his wood lot to orchard grass he can have a pasture equal to the blue grass pastures of Kentucky. In a rich soil the china tree has been known to grow from seven to ten inches in diameter in three years from the seed. The limbs grow out ten or twelve feet in a single year, and as large as a man's arm; and they burn well even when green. The wood has a firm and fine grain; it makes splendid cabinet work, and is durable as fence posts. No worm or insect will live on or near it, and it makes a beautiful shade tree.

THE SALT MINE.

Petit Anse Island lies one hundred miles west of New Orleans, and three or four miles north of Vermilion Bay, in the parish of Iberia. It has a surface of about 2400 acres. It rises up a mountain in the sea marsh, its summit being 185 feet above the level of the Gulf. It contains a mine of pure salt, on the south side, in the lowest places but fifteen or twenty feet below the surface of the land, the surface of the mine being about level with the waters of the Gulf.

This mine is known to be half a mile square; its depth is unknown. A shaft has been sunk into the mine to the depth of seventy or eighty feet below the surface of the ground, if we remember correctly. It is about six feet square, and resembles a plank-curbed well. An elevator is lowered into the mine and raised out of it up and down this shaft, by a stationary engine.

A chamber has been worked in the mine 7 feet high, 27 feet wide and 300 or 400 feet east and west from the bottom of the shaft. The miners drill holes in the solid salt, which appears like marble, and blast it into innumerable shapes, and masses, as they blast rocks. A car, on rails laid for the purpose, takes the salt from the end of the chamber to the shaft, where it is raised to the surface above, crushed in mills by steam power and put in marketable shape. This mine has not been worked for some years.

Petit Anse Island, now owned by Judge Avery, has a rich soil, adapted to all productions of Southern Louisiana. It has been chiefly devoted to sugar culture, but it is adapted to sugar, cotton, rice, tobacco, and all kinds of vegetables.

THE SULPHUR MINE.

About 80 miles west of Petit Anse, or 180 west of New Orleans, and 30 miles from the open Gulf, on the west fork of the Calcasieu river, near Lake Charles, is an immense mine of pure crystaline sulphur. The surface of the mine is 428 feet below the surface of the soil above it. The strata of pure sulphur is a hundred feet in thickness, and below this, to the depth of 700 feet, there is a mixture of sulphur and gypsum, mostly gypsum.

Heavy expenses have been incurred by the Sulphur Mining Company in making preparations to work this mine, but thus far they have been unsuccessful. Machinery for this purpose, even the iron curbing, was brought from France. They attempted to make a circular shaft 12 feet in diameter, and to curb it with iron like boiler iron, put down in rings or sections, as they proceeded with the digging. The sections were to be riveted together like the sections of a steam boiler, and made water-tight.

PETROLEUM.

Many years ago petroleum oozed out of the ground near Lake Charles in considerable quantities, and in boring for oil they struck the sulphur mine. They have never yet struck the desired fountain of oil, but will doubtless find it at some future day, in the southwestern part of the State, in large supplies. But this mine is more particularly described in our chapter on Calcasieu.

RICE FIELDS OF LOUISIANA—HAY AND GRASS.

The following letters were written by the author of this book, in September, 1872, and published in the N. O. Times:

THE RICE FIELDS OF THE LAFOURCHE.

Rice culture is becoming an important branch of agriculture on the Bayou Lafourche and Morgan's Louisiana and Texas Railroad. The profits of rice culture have this season been quite satisfactory—1873.

THE RICE PLANTING SEASON.

The rice planting season extends through March and April into May. One man can plow and sow about twenty-five acres of flooded rice. It is usually sowed in drills, and the grass and weeds which the flooding fails to kill, are worked out by hand and hoe. The rice crop of Lafourche parish is about three-quarters of it cultivated by Creoles, and the balance by negroes. The white men who work in this crop are not more sickly than the negroes. Some who work this crop think it not quite as healthy for the hands as the sugar crop. Others say they can see no difference between laborers in rice and sugar fields in these respects.

THE RICE HARVEST.

The rice harvest is usually completed in August. This crop matures in four months or less. Harvesting requires about twenty-six days, to reap, bundle, and stack the crop. A man can reap from a quarter to half an acre daily. They expect soon to apply reaping machines to rice harvesting, The softness of the ground has been the chief difficulty in using reaping machines with horses and wheels.

YIELD AND PROFITS.

It has usually been said that an acre will produce ten barrels o flooded rice, or seven barrels not flooded. The Lafourche rice cultivators say that seven barrels are a good average yield per acre for a flooded crop.

A barrel of clean rice is 230 lbs.—it usually sells for 7c. a lb. —an acre would amount to........................	$ 112 70
To plow, sow, weed, hoe and harvest an acre of rice costs the farmer about....................................	28 00
	$ 84 70
25 acres of rice at $112 70 an aere, would amount to........	2,817 50
The hired help in making and saving 25 acres, the crop for one hand, would not amount to more than $12 or $14 an acre, say	350 00
	$2,467 50

Threshing machines that turn out 60 barrels or more daily, at a small expense, are used for separating the rough riee from the straw.

LOUISIANA RICE CULTURE—GENERAL FACTS.

In leasing rice lands on the Lafourche, the landholder has one-fourth of the crop and the laborer three-fourths. The laborer furnishes his own team, plows, seed, etc., and pays all expenses.

The lessee of rice lands can plow and sow about twenty-five acres of land without hiring any help. He has to hire by the day, usually $1 50 per day, to help weed and harvest the crop.

He may calculate, one year with another, the following results :

25 acres rice, 7 bbls. to the acre	175 bbls.
One-quarter deducted for lessee's share	43¾ bbls.
Lessee's share	131¼ bbls.
131¼ barrels, 230 lbs. to a barrel, at 7c. a pound	$2,142 35

An enterprising farmer who owns the land he cultivates, with a a few boys, may plant and cultivate a hundred acres of rice and not pay out over $14 an acre for extra help.

100 acres of rice, 7 barrels, 230 lbs. each, 161,000 lbs., at 7c.	$11,270
Extra help in cultivating and harvesting, $14 an acre	1,400
	$9,870
Cost of cleaning at rice mill, 1c. a pound	1,610
	$8,260

After paying freights, commissions, storage, etc., there is still a handsome margin for the farmer.

Then all this is done in about five months, leaving seven months, or six at least, to be devoted to other labor.

And in this we are reckoning but seven barrels of clear rice to the acre, while many rice planters have claimed that the average yield is ten barrels to the acre, or 2300 pounds.

RICE CULTURE IN ST. MARY.

The Sypher Brothers, on the Chattworth Plantation, near Franklin, Parish of St. Mary, last season cultivated 80 acres of rice. They used the Andrews' Centrifugal Pump in flooding the rice field. The pump raises over their low embankment 5,000 gallons a minute, and increased the depth of water one inch in 24 hours in the field of 80 acres.

They sowed the seed rice on the 25th of May, and harvested the crop about the 25th of August. They used grain reapers, altered to suit rice, in harvesting the crop. They used slides with the wheels, and reaped eight to ten acres daily. They made from twelve to fifteen barrels of rough rice to the acre. Rough rice weighs 162 pounds to the barrel—clean rice, 232 pounds. They are satisfied with the results, and will plant 150 acres of rice next year.

HEAVY RICE YIELD IN ST. CHARLES.

One of our active Creole rice planters of the parish of St. Charles, Theophane Thiel, Esq., is getting twenty barrels of rough rice to the acre from his fields. This gentleman has a rice field of one hundred and forty acres, which he works scientifically. This is a yield of about 66 bushels to the acre, which is a very heavy one as compared to the general yields of our rice fields under the present mode of cultivation practiced here.

Under a proper system of cultivation, every acre applied in the State might be made to produce this amount annually.

This gentleman shipped to the city, on the 21st, the product of three acres, 60 barrels, which, after cleaning by the "Brook Steam Rice Mill," sold for $404 75.

These figures were shown us on the books of our active young rice merchants, the Thompson Brothers. An acre of wheat returns about $25 gross product. This is a return for one acre of rice, $135. Rice culture is destined to be an important interest in Louisiana.—*New Orleans Times.*

SUGGESTIONS TO RICE PLANTERS.

In a trip through the rice regions above the city and on the Lafourche, we observe a good second crop in many fields, sprung from the old stubble cut two months ago. This crop is ripe, or ripening very fast, and will make enough in many places to pay all the expenses of the first crop, thus leaving the original crop clear profit. With a proper system of irrigation, so that water can be raised from the streams and lakes, by wheels or by pumping, the second crop of rice could be always a certain crop in lower Louisiana.

At present the only rice fields which are cultivated in the lowland rice, which requires irrigation, are situated along the Mississippi river and its outlets or bayous, or, as the Spaniards call them more correctly, *riachos*.

The rice plantations depend upon the rise in the Mississippi for the water for their rice lands. When the river is above the level of these lands the water can be let into the fields through gates in the levee. This is a precarious system and occasionally cuts short the rice crop when the low water prevents the flow into the fields from the natural level.

As rice culture is becoming one of our greatest agricultural interests, nothing should be left to chance when it is possible by moderate expenditure to make success certain.

The simplest contrivances for raising water are current wheels, or floats or flats. Windmills are also a cheap and easily managed power for pumping, and they are now self-adjusting and self-reefing, so as to protect themselves against sudden flaws or storms.

But steam power is now so very cheap that every body may have permanent engines and boilers put up at a very low figure.

The question of fuel, which makes so alarming an item of cost in the estimate of running an engine, need not trouble the rice planter. It has been found in California and some parts of the West, as well as on the continent of Europe, that steam threshing machines can be carried into the fields and run by threshed straw. The rice straw is in the planter's way now, but he could pump his water readily if boilers and furnaces were set to use it, as it is a strong fuel, and could use his engine for threshing and various other purposes.

We bélieve, now that a more intelligent class of planters are embarking in rice culture, that many improvements will be made, not only in the modes of culture and irrigation, but in the introduction of many labor-saving machines.—*Co-operative News.*

RICE.

ITS PRODUCTION RAPIDLY INCREASING—A VERY RICH AND PROFITABLE INDUSTRY—THE PRESENT CROP AND PROSPECTS.

The growing importance of the rice production of the current year will surprise even the most sanguine advocates of the cultivation of this staple. In cértain portions of St. James parish, as many as twenty-two barrels of rough rice have been produced upon an acre, and the crop of 1875 is estimated at four hundred thousand barrels. Heretofore the net profits per acre after paying all expenses, including cultivation, etc,, has been $47 50, but this was upon an estimate of twelve barrels per acre. Should the amounts already reported prove a criterion, the profits would be increased by at least eight barrels per acre, less the cost of milling and freight. The great advantage is the small cost which attends its cultivation. Quite recently several hundred colored laborers were secured in the city for the purpose of harvesting the crop, it being ascertained that one man could plant and cultivate more than five could gather. The labor entailed in this is indeed trifling. The ground is turned to the depth of but a few inches, and after the rice has sprouted, irrigated by a sluice gate from the river, the water is permitted to cover it several weeks, and then drained off. A second irrigation and drainage completes the task. But a small investment is necessary until the crop is ready for harvest, and then advances sufficient to secure all the necessary labor, can readily be procured. Not only is it expected that the present crop will yield a year's profit of nearly six millions of dollars, but a sufficient supply will remain in the country to provide laborers with abundance of food. The production of the State will approach ninety-two millions of pounds to furnish over two pounds to every human being in the nation. Producers of this staple have wisely concluded not to undertake too much, and in consequence its milling and polishing has become a separate industry. Hundreds of thousands of dollars are invested in the latter enterprise, and while a fair profit is made by millers, they can afford to purchase the product for market cheaper than could the planter, were he to erect his own machinery.—*N. O. Times.*

FRUIT CULTURE IN LOUISIANA.

TROPICAL AND SEMI-TROPICAL FRUITS.

Oranges — Lemons — Bananas — Limes—Shaddocks — Citrons—Figs—Olives—Dates—Sago Palms—Guava—Jujube.

The above list of fruits can all be raised in Louisiana, most of them as high as 30°, some far above; others can hardly be raised above latitude 29° 30′.

ORANGES—THE ORANGE CROP.

It is estimated that last year the orange crop of Louisiana amounted to 65,000 barrels. It is difficult to obtain a true statement of the crop raised in the State. The small groves of orange trees, scattered all over Southern Louisiana, near and below latitude 30°, from the Sabine to Pearl river, on numerous bayous and lakes in and near the sea marsh, five, ten, twenty, or fifty in a grove, and individual trees planted near dwelling houses, cannot be summed up and given to the reader in acres with any accuracy.

In establishing orange orchards, from 100 to 125 trees are usually planted on an acre. They are planted from fifteen to twenty feet apart. Probably there are from 1000 to 1500 acres of orange trees in the State. The home consumption of oranges that never reach the market is immense.

Oranges are raised in great perfection in the parishes of St. Bernard, Plaquemines, Orleans, Jefferson, St. Charles, St. John Baptist, St. James, Assumption, Lafourche, Terrebonne, St. Mary, Iberia, Vermilion, Cameron and Sabine. They are also raised in favorable localities in the lower parts of several parishes above the 30th parallel. There is orange land enough in Louisiana to increase the orange crop of the State a thousand fold and to supply the whole Mississippi Valley with oranges. Most of the sea marsh of Louisiana, if leveed and drained, as the lands of the Netherlands have been reclaimed, would be admirably adapted to orange culture.

Oranges this year (1875), have been raised on the N. O. St. L. & C. Railroad, as high up as Independence, in the pine woods, 65 miles, or nearly one degree above New Orleans. The trees are vigorous and healthy, and the fruit fine. Several bushels have been raised from a single young tree. More attention will be given to oranges in the pine lands in that section of the State, though the latitude is a little too high.

LARGE ORANGES FROM TREES SIXTY YEARS OLD.

Editors of Our Home Journal:

I send you a few oranges to show you what we can do in this part of the country. They are from trees sixty years of age, planted by my grandfather; standing to-day where the seeds were planted, without having ever been removed. The trees are perfectly healthy and full of fruit.

Yours, respectfully,

F. B. HUDSON.

Bayou Salé, parish of St. Mary, La. Nov. 20, 1875.

The editor of the Home Journal remarks:

We received the oranges all in excellent condition, and return thanks to our friend. Every one who saw them pronounced them not only the finest, but the largest they had ever seen. The box, the size of a sixty pound soap box, was filled with twenty-nine oranges, the largest weighing 17 ounces, and measuring 13 inches in circumference, and the smallest 10 ounces—the twenty-nine oranges weighing 23 lbs. 12½ ounces, or an average of 13 1-10 ounces each, and every one of them was a perfect beauty. This excellent sample from the rich lands of the Teche country, gives a faint idea of what is in store for Louisiana in the future, when her now thousands of acres of idle land will be converted into rich orange groves, with crops worth $1000 an acre.

Trees sixty years old, and such fruit! Never transplanted, and perfectly healthy! What a fortune there is in store for those who now plant trees and give them careful and scientific culture. Is there any crop that will pay better, and that will give a lasting income like it?

DATES.

Dates have succeeded in Southern Louisiana in the open air, but few have ever given them any attention.

SAGO PALM.

The sago palm grows well in the coast parishes, and even on the river above New Orleans. They are chiefly cultivated in connection with ornamental trees and shrubbery. They sometimes grow as high as thirty feet, and are admired for their rank tropical character and appearance.

JUJUBE.

The jujube is a thorny shrub, hardy, and easy of propagation. It grows in Lower Louisiana, and appears to be hardy enough for a hedge plant, or to grow in higher latitudes.

GUAVA.

The guava succeeds in the coast parishes of Louisiana, and perhaps in higher latitudes. It is seldom cultivated in this State.

MESPILUS OR JAPAN PLUM.

This is a fine plum, yellow, of a pleasant acid, grows on a beautiful evergreen tree; the fruit ripens in March or April, blossoms in the fall, and is not injured in the winter by moderate frosts below latitude 30°. The tree will grow one or two degrees farther north, but the fruit does not always mature in the vicinity of New Orleans. There was plenty of this fruit ripe in the New Orleans market this year by the 1st of February.

LEMONS.

Lemons are raised in the open air and without protection as high up as 30°, the latitude of New Orleans; but they are not a sure crop above latitude 29°½. We have seen them ripen in St. Mary's Parish in latitude 29° 30'.

BANANAS.

Bananas ripen in the open air in the latitude of New Orleans, in positions favorably situated in regard to frosts. We have known them to ripen on Bayou Teche as high up as Franklin. After a very mild winter they will ripen anywhere as high up as latitude 30°, and perhaps a little above.

LIMES.

Limes are not so hardly as lemons. We have never seen them cultivated successfully quite so high up as New Orleans. They can be raised near the coast, and on the islands where water on the north side protects them from injury by frosts.

CITRONS.

Citrons grow on a low shrub, and stand nearly as much cold weather as the orange. They grow in great perfection in the southern parts of the Parish of St. Mary, and in Terrebonne, and other coast parishes

SHADDOCKS.

Shaddocks do well in the southern part of St. Mary, and in all of the coast parishes.

OLIVES.

Olives succeed well in the coast parishes, and would do well doubtless as high up as the northern part of the State. But few have ever been cultivated even in the coast parishes, but those few have done well.

ALMONDS.

Col. S. J. Matthews, the great Arkansas nurseryman, sent us the other day, a box of almonds grown upon his place near Monticello. This was a pleasant surprise, although we had before seen an announcement that he was developing the almond culture. They were much more palatable than those brought from a distance; were large in size and perfectly framed.—*Arkansas Paper.*

FIGS.

Figs grow in all parts of Louisiana, but there are some varieties that are too tender for a latitude much above 30°. Every morning in the fruit season, delicious fruit that ripened during the night can be taken from the trees.

Most kinds of apples do not succeed well in any part of this State. There are some kinds, however, that do very well in the middle and upper parishes, and some that, with skillful management, have done well as low down as latitude 30 degrees.

Mr. H. W. L. Lewis, Master of the Louisiana State Grange, has succeeded in several varieties, raising fine apples. He has experimented with over fifty varieties and about the same number of varieties of pears and grapes. He has a seedling Southern apples that he thinks a complete success. He finds but few trees out of those he has experimented with that he thinks worthy of much attention. His orchard and nurseries are in Tangipahoa parish, five or six miles from the State line, and near Osyka, Mississippi.

PEARS.

Pears do exceedingly well in various localities, from the shore washed by the waters of the Gulf of Mexico to the upper line of the State. Magnificent pears have been raised on the banks of the Teche, and in various places in Attakapas and St. Landry there are a few fine healthy bearing pear trees. No better pears are found in the State, if in the South, than the best qualities of this fruit produced within sound of the waves that roll upon the beach between New Orleans and Mobile. Pears require a soil that has considerable iron in it. The climate of Louisiana is fine for pears in all parts of the State, and science and experience will doubtless be able to correct the defects of the soil in varions localities, and pears may become a profitable crop in this State.

GRAPES.

The scuppernong is doubtless the most valuable grape in the South. It succeeds well from the gulf shore to the Ohio River, and perhaps beyond. It does well in either alluvial or sandy, pine woods soil. Give it a good arbor, increase the arbor as the vine extends from year to year, give direction to the new shoots and do not allow the vine to grow in bunches, and in almost any soil properly drained and not too rich, a single vine will make fruit and even wine enough for a family.

These vines that produce such extraordinary crops have been planted near old wells, and the roots run down and around the decayed curbing to the surface of the water. The well furnishing underdraining and air more completely than can be done by any other arrangement. The vine lives and thrives to the age of two hundred years or more.

The Concord and Ives seedling succeed admirably in the pine lands of this State. There are several other kinds of grapes that do well. The Warren, Diana, Tennessee, Delaware, Hartford, Prolific, Roger, Red River, and some of the native grapes of Louisiana are highly prized. The Red River grape, now cultivated in France, is a native grape of this State. The muscadine grows in all parts of the State, and, like the scuppernong, grows in any kind of soil, from swamps to pine hills. The pine lands are the best soil for grapes.

THAT WONDERFUL SCUPPERNONG.

The following article from the *Rural Carolinian*, will illustrate the prolific qualities of this valuable grape:

In your April number I see that you allude to a scuppernong vine grown by me. I do not know how you received your information relative to the marvelous growth of this vine. But justice demands for this generous vine that I give the facts in the premises. In June of 1867, I layered a few vines of the Scuppernong in a lot I owned in Jacksonville. In July of the same year, I sold this lot to Mr. C. L. Robinson, reserving these vines. In the fall of 1867 I planted these vines, one of which was near the well; this vine covers a lattice-work fifty-four by sixty-four, and will bear this season one hundred bushels

I think. This vine can be seen without a search warrant, and I can prove the above facts.—*Florida Correspondent.*

SWAMPS.

Plums succeed well in all parts of this State. But little efforts have been made to introduce new and improved varieties, but some of the varieties raised are quite large and very good fruit. The plum, like the fig tree, is expected to take care of itself, and but little attention is given to it only to keep the cattle from destroying the trees when young. In fact, the plum, the orange and the fig are the only fruits that have been produced in the gardens and around the dwellings in Southern Louisiana for home supplies, with the exception of a few strawberry patches and a few grapes and pears, for the reason that these fruits, in former years, required but little attention. These and blackberries and dewberries, which grow bountifully all over Southern Louisiana, constituted the chief fruits of this section of the State.

PEACHES.

But little attention has been given to peaches, for the reason that they require more attention than plums, oranges, figs and grapes, and they are usually allowed to be smothered with weeds, choked with grass or destroyed by cattle. When our people attend to peaches as they do in Delaware and New Jersey, they will have an abundance of the fruit in all parts of the State. Peaches do better in the middle and upper part of the State than in the lower part. Formerly, peach orchards were much better attended to in all parts of Louisiana than of late years, and good peaches were raised in abundance clear down to the Gulf shore, and on the islands. Good cultivation and proper attention would doubtless again give us good peach crops in this State.

APRICOTS AND NECTARINES.

No attention has been given to apricots and nectarines, but a few persons have succeed well with them. Like peaches, they will doubtless succeed when properly cultivated and cared for by skillful managers.

QUINCES.

The common quince does not succeed well in Louisiana, but perhaps it is as much for want of proper culture and attention as anything else. The Chinese quince produces well every year in the coast parishes, and the tree and fruit are healthy. It is superior to the common quince.

STRAWBERRIES.

Strawberries do well in all parts of the State, from the Gulf shore to the upper parishes, when properly cultivated and when the soil is properly prepared.

BLACKBERRIES, DEWBERRIES, ETC.

Blackberries grow bountifully in all parts of this State. They might be dried and preserved, or made into wine, and become a valuable article of commerce, as in North Carolina and other States. Dewberries grow wild everywhere in the State. The vines are hardy and prolific, and the fruit excellent.

CONCLUSION.

Louisiana should be a great fruit growing State, and should send to market yearly millions of dollars worth of ripe, canned, preserved and dried fruits and wines. We need an immigration of a large number of enthusiastic and practical fruit men to this State. We hope to see them here in a few years.

PEARS ON BAYOU TECHE.

In July, 1873. we visited the orchard of Capt. Eugene B. Olivier, on Bayou Teche, parish of Iberia, ten or twelve miles from the Gulf coast and about 120 miles west of New Orleans. We there wrote the following letter, which was published in the N. O. Times on the 4th of July:

A short time since we went through the pear orchard of Capt. Eugene B. Olivier, situated on the west bank of the Bayou Teche, five miles below New Iberia, in the parish of Iberia. Several of his trees were hanging so full of fruit that the limbs were breaking under the weight. We counted about a hundred fine pears on two branches, and, though the tree is but half grown, it has not less than a thousand pears on it. Other trees were groaning under the weight of as perfect fruit as grown in any latitude. Capt. Olivier has about 100 bearing trees, many of them twenty years old, besides many young trees. He has seven varieties of pear trees that deserve special notice, the Fondante d'Eté, Louise Bonne, Soldat Leboureur, Beurre Admirable, Beurre

Goubault, Bartlett, and St. Michael Archangel. They ripen in July and August.

AN INTERESTING LETTER FROM CAPT. OLIVIER

The following letter was written by Capt. Eugene B. Olivier, two years later, and published in the SUGAR BOWL, at New Iberia. Capt. Olivier is excellent authority on the Louisiana fruit question. We know of no better authority in the State. We have been intimately acquainted with him for more than a quarter of a century, and freely vouch for the truthfulness of his statements, and his ability and intelligence in connection with fruit culture. Any one who has partaken of his hospitalities, and scrutinized his orchards, fruits and nurseries, as we have done, will always remember him as a fine specimen of the intelligent creole gentlemen of former days, and as an enthusiast in the fruit business; and a *successful* enthusiast.

PARISH OF IBERIA, July 25, 1875.

Editor of Sugar Bowl:—Accompanying this, please find samples of five different varieties of apples, submitted to your appreciation *de visu* and *de gustibus*, and after scrupulous analysis, please tell your readers what you think of Louisiana apples, or rather of the *possibility* of raising apples in Louisiana, which to most people remains as yet an open question. The sample, of which you'll find one wrapped in paper, is a native variety, found in East Baton Rouge, and much spoken of some three years ago.

Besides these varieties, several others, of earlier and later maturity, do equally as well here with half the care the trees should receive.

Nowhere does nature offer a more admirable locality for the culture of fruits than those fine undulating plains, stretching from New Iberia westward; and yet, what efforts are being made in that healthful and remunerative industry? People of Iberia, St. Martin, Vermilion, Lafayette, St. Landry and Calcasieu, awake to the voice of nature, which calls you in unwavering signs, to the undertaking; plant the peach, the apricot, the nectarine, the fig, the pear, the apple, the orange, the grape vine—can your own fruits and make your own wines. If you don't—if you disdain to reap the treasures strewn at your feet—others in due time most certainly will, because nowhere was there ever more pointed indications of the destiny of your country. Therefore, ye of Calcasieu, drive off your liliputian cows and knock-kneed ponies, and let the grape vine wave its treasures throughout your broad prairies; dot your water courses with the beautiful orange tree; and ye of the intervening region, let us have an abundance of the fruits named above; acclimate others, cultivate none but the very best varieties, create new ones, and "hide not your lights under a bushel," but, *pro bono publico*, tell your neighbors of your experience—how you stumbled and got up again; of the many blunders committed, and the success encountered, as a means of mutual benefit.

GRAFTED ORANGE TREES.

Of late years considerable attention is being given to grafting orange trees. A seed planted in the winter, or spring, in good soil, and properly attended to, may make a stock large enough to graft, or bud the next spring, when one year old, but a two year old stock is better. In two or three years the graft will produce fruit. A seedling orange tree often requires from 8 to 12 years to come in bearing.

They usually graft on the sour orange stock which is considered more hardy than the sweet orange tree, and less liable to a certain disease which attacks the bark of the tree at the surface of the ground; the sap disease.

The Nelson Brothers, at the Magnolia Nurseries, in the suburbs of New Orleans, are grafting the Brazilian orange scion on the sour orange stock, also the Mandarin, and the Creole orange. They consider the Brazilian, in some respects, superior to the Creole orange, and the Creole as much superior to the Cuban, or the Mediterranean oranges. The Nelson Brothers are reliable gentlemen, and will give correct information to inquiries in regard to orange culture in this State.

THE DEMAND FOR LOUISIANA ORANGES.

The demand for Louisiana oranges always keeps ahead of the supply. As railroad and other transportation facilities increase, the great West calls for a larger yearly supply of this delicious fruit. Louisiana can never glut the orange market of the great valley of the Mississippi. The better portions of the sea marsh, on the coast of Louisiana, will doubtless in time be leveed and planted in orange groves.

VARIETIES.

There are about ten varieties of oranges now raised in Louisiana, the principal varieties are the sweet, sour, myrtle, Mandarin, and Brazilian. The sour orange makes magnificent preserves, and the juice a delicious sour orange wine.

The Brazilian orange is delicious, not quite as acid as the Creole, hangs on the tree later in the season than the creole, is hardy, prolific, and does well grafted on the sour orange stock.

The Mandarin is a delicious little orange, can be pealed without a knife, and divided into ten or a dozen sections, without starting the juice. A lady may peal and divide it without soiling her gloves.

The myrtle orange grows in rich clusters, and the trees are sometimes buried in thousands of its golden fruit. The tree, and leaves, and fruit, ore much smaller than the Creole orange. The fruit is only used for making preserves.

The bastard orange is a medium between the sweet and sour orange; rarely cultivated

FRUITS FROM OUR GULF COAST.

From Our Home Journal, Sept. 18, 1875.

At the recent meeting of the American Pomological Society, in Chicago, Mr. Redmond, of Ocean Springs, Mississippi, exhibited the following fruits from our section :

Kinds.	No. of Varieties.
Oranges	9
Lemons	2
Bananas	2
Pine Apple	1
Date	1
"Custard Apple," (carica papaya)	1
"May Apple," (passiflora edulis)	1
Jujube	1
Figs	4
Grapes, (Southern sorts)	2
Chinese Quince	1
Pears	1
Apples	2
Persimmons	1
Pomogranates	4
Olive	1
Of all varieties	37

Mr. Redmond desires to return special thanks to Messrs. Alfred E. Lewis and F. Gautier, of Pascagoula; to Mr. U. Cousins and Mrs. Sarah A. Dorsey, of the Coast; to Brother John, of the College at Pass Christian; and to Messrs. Sambola, Devron, Rountre and Joseph Muller of New Orleans, for very efficient aid in making up his collection.

FIGS.

There are twelve or fifteen varieties of figs cultivated in Louisiana, but the best and most reliable of all is the cleste or sugar fig.

The blue or purple fig is large and fine, particularly fine in a dry season. None of the varieties of figs are as good in showery as in settled, fair weather.

The Marseilles is a splendid fig, particularly fine for preserves, but the tree is not as hardy as the other kinds.

The red Mexican fig is larger and later than any others, but it is coarse, and is not as sweet as other kinds.

The old, common yellow fig, most usually growing in the country is hardy, and a good fig in dry weather, but of little account in showery weather.

Most all kinds of figs are prolific, and bear one good crop every year. The early crop is generally worthless. The second crop is the only one of value. The third crop, in the latter part of summer, is better than the first, but it seldom amounts to anything, so far as our observation extends. The figs are neither as sweet or abundant as the second crop, and do not mature as perfectly.

The fig tree and fruit have no enemies in the form of insects, bugs, worms, or diseases. It is as hardy in its true latitude as a live oak, only the young trees sometimes suffer a little in a hard winter.

THE VALUE OF THE FIG CROP.

The fig crop, like the orange crop, should become exceedingly valuable to this State. The yearly crops are bountiful. The yield of an acre of fig trees, properly managed, in full bearing, would be enormous. They live and bear fruit until they are fifty years old, or older.

Millions of jars of preserved figs should be sent yearly from this State to the North and West. They cán be put up in a marketable shape in numerous ways. The finest pickles we ever saw were made of figs. Ripe figs make fine vinegar. From the beer prepared from ripe fruit, doubtless fig brandy could be made of an excellent quality, far better than peach or apple brandy. More than three-quarters of the State of Louisiana is adapted to fig culture. They would grow and produce bountifully on reclaimed swamps, and sea marshes.

WHEAT—CORN—COTTON—BREAD AND MEAT.

WHEAT.

The yield of wheat in Ohio, fifty years ago, was 30 bushels to the acre; to-day it is not quite 15.

In 1775, the wheat fields near Albany, N. Y., yielded from 30 to 40 bushels to the acre, with inferior cultivation. In 1855, winter wheat yielded 7½ bushels per acre, and spring wheat 5 bushels.

In Maine, wheat has nearly ceased to be cultivated where formerly good crops were made.

In Mississippi, and in some of the upper parishes of Louisiana, farmers have this year, (1875) made in many instances from 20 to 30 bushels of wheat to the acre, superior in quality to any made north of the Ohio river or in the Great West. Southern wheat always brings a higher price in any wheat market than Northern wheat.

CORN AND MEAT.

"After thirty-five years of planting in Louisiana," said Mr. C. yesterday, "the result of my experience is that the man who plants corn and makes his own meat and bread, will have money to loan at the end of the year, while he that raises cotton alone, buying his meat and bread, will be compelled to borrow money."—N. O. Times, Mch 30, '75.

CHEAP CORN.

A farmer on the New Orleans, St. Louis & Chicago R. R. made two hundred bushels of corn with twenty-five days work, on land not considered the best. The same may be done every year on lands properly manured and well cultivated. Why should Louisianians bring corn from 600 to 1500 miles, or more, at heavy expense, when a day's work may secure eight bushels of the article of better quality than Western corn?

CORN AND WHEAT.

A Western farmer says: "The corn crop does not average 40 bushels to the acre in the West; some make 50 to 60 bushels, others from 10 to 15."

Wheat sometimes yields 25 to 30 bushels, but oftener from 12 to 15 bushels.

An acre of corn in the crib is generally worth about $12.

A good farmer in Louisiana, who manures his lands as he should do, can get from 25 to 30 bushels to the acre, and more than that amount of oats. Farmers in Mississippi have in many instances made 25 bushels of wheat to the acre this year; some have made more.

A BALE OF COTTON TO THE ACRE.

A successful farmer in the pinelands informed us that he has made four bales of cotton, 500 pounds net each, on four acres of pine land, by applying a moderate amount of manure to the land. He thinks that all pine lands that have a good subsoil can be easily made to yield 35 bushels of corn, or 500 pounds of lint cotton to the acre.

In Georgia five bales of cotton have been made on a single acre in one year.

CALCULATION OF THE PRICE OF A COTTON CROP.

If the figures 60 are divided by the cotton crop in millions of bales, the result will give the average price per pound at the nearest market. Thus the crop raised is

2 million bales	60	30 cents	per pound.
3 ..	60	20 ..	..
4 ..	60	15 ..	..
5 ..	60	12 ..	..

It will thus be seen that under ordinary circumstances, three million bales of cotton are as valuable to the planter, and will bring as much money as can be obtained for four or even five million bales.

A small crop is increased in value by the necessities of the consumers. A large crop is reduced in value by any disturbance of the money market, or by war, pestilence or famine. The surplus land can be profitably occupied in producing pork, corn and hay, without reducing the value of the cotton crop one dollar.

The more corn, pork, hay and provisions made in the South, the higher will be the price of cotton, one year with another. The South has the physical force now in the field sufficient to make its bread and meat, and at the same time make a cotton crop that will yield a larger aggregate profit than 4,000,000 bales.

CAN WHITE MEN WORK THE SOIL AND STAND THE CLIMATE OF LOUISIANA?

MR. PETER PECOT'S STATEMENT—PROFITABLE FARMING.

Mr. E. Maynard, in the year 1869, a small farmer in the Parish of St. Mary, by his own labor, and hired labor that cost him $145, made enough sugar cane to yield forty-five hogsheads of sugar and sixty barrels of molasses, The net proceeds of his own labor, after paying for all hired labor, amounted to four thousand four hundred and forty dollars.

The next year, assisted by his son and nephew, fourteen and sixteen years of age, and six hundred days' work hired in sugar making, at times when extra work was needed in the crop, made 57 hogsheads of sugar and 75 barrels of molasses. He also made his home supply of corn, sweet and Irish potatoes, rice, forage, etc. After paying all of his farm and family expenses, he had $4800 profits left as the result of his year's work. Mr. Maynard is a Creole, of French descent, a native of Louisiana. His crops were made below latitude 30°.

SUGAR CULTURE BY WHITE MEN.

On Bayou Teche, Parish of St. Mary, in 1871, there was a colony of one hundred white persons on M'me Grevemberg's plantation, cultivating sugar cane on the share system, twenty-five hands in the field.

They each cultivated about 7 acres of cane and 13 acres of corn, and cultivated their crops better than any worked by colored laborers.

The books of the plantation show that the following amounts were paid to different hands as their share of the crop aftér it was sold in the New Orleans market:

Paid over to B. Delotte	$ 659 23
Paid over to Taylor Delotte	536 61
Paid over to A. Thibodaux and little boy	575 00
Paid over to Guillot Gerazime	475 00
Paid over to J. B. Ilert and 2 sons	1,462 37

Paid over to Gustave Rogers (an old man)	971 50
Paid over to 3 young men, Dupuy Brothers	1,728 16
Paid over to a man about 50 years old	450 00
Paid over to an old man and 2 sons, 20 and 17 years, the old man working but little	1,316 06
Paid over to a man and 2 sons, commenced late in the season	1,643 44

Three men in a family made 32 hogsheads of sugar, and about 50 barrels of molasses.

The men all used their own horses and teams; not a negro in the colony in house or field.

The doctor's bill of the entire colony did not exceed $150 a year. Sometimes not over $100, as stated by their physician; $1 to $1 50 a head, and the physician lives two miles off.

Each family made its own corn, potatoes, etc., and sold from $100 to $150 worth of eggs yearly.

Net proceeds of sugar and molasses made by the colonists in two years, $40,000, half for the planter and half for the colonists.

MR. WM. HUDSON'S STATEMENT.

Mr. Hylaire Hebert came to Mr. Hudson's farm in 1871, with a large family, in great distress. He had no property or money. Mr. Hudson is a merchant, living at Jeannerette, on Bayou Teche, Parish of Iberia, and has a small farm near by. He employed Mr. Hebert and his family to work his farm on shares. He furnished five arpents of seed cane, and Mr. Hebert planted 23 arpents of cane and plenty of corn. In the fall he and his boys made 37 hogsheads of sugar and about 50 barrels of molasses, put up seed cane for next year, and put up 175 barrels of corn for his share.

In 1872, he, with the same force, made 33 hogsheads of sugar, about 50 barrels molasses, (molasses about 60 gallons to a hogshead of sugar), and plenty of corn and forage. The first year he got $1500 cash for his part of the crop, the second year $1700. He sold eggs, chickens and produce, had but little sickness in his family, and laid up a handsome sum of money. Mr. Hudson made, clear of expenses, 22 per cent. on the value of his farm the first year and 16 per cent. the second year.

MR. PATOUT'S STATEMENT—1871.

Mr. Patout, who lives in the Au Large Prairie, Parish of St. Mary, La., and works a plantation near Jeannerette, says that foreigners and Northern men can stand field labor in this country as well as natives of the State, white or black. He is working white laborers on his plantation on the share system and by the month. During slavery his doctor's bill for one hundred slaves and ten whites did not exceed a hundred dollars a year.

Yellow fever has never been epidemic in the Au Large Prairie.

He thinks that one good hand and two mules could cultivate forty acres of corn on his plantation, all but thinning out.

MR. THIBODAUX'S STATEMENT—1871.

Mr. Thibodaux worked on M'me Grevemberg's plantation three years, cultivating cane on the share system. He had a wife and ten children, seven of them boys. He made enough in three years on shares to buy a farm of thirty-five acres, with a good house and barn. Had a good crop of cane, corn and rice. His family had more sickness than any in the colony. He had the pleurisy, which cost him $35 for doctor's bill. He had paid only $155 for doctor's bill for himself and family in six years.

STATEMENT OF MR. PATOUT AND THE CATHOLIC PRIEST.

In the Au Large and Cypremort prairies, in the parishes of Iberia and St. Mary, west of Bayou Teche, in latitude 30 degrees, near the coast of the Gulf of Mexico, in a distance of seventeen miles, and a width of five miles, there are (1871) three hundred small farms, worked and owned by white farmers and their sons and hired men. But few negroes employed. The farmers' wives and daughters usually do their own housework, cooking and washing. There are also about seventy-five white families working on shares in the same section of country. They are more healthy than the negroes, and live longer, they work in the field in all seasons of the year, and do all kinds of work that negro field hands do in the sugar regions. They are much better farmers than negroes.

WHITE MEN IN COTTON AND CANE FIELDS.

TESTIMONY OF INTELLIGENT SOUTHERN FARMERS.

A circular published in 1869, indorsed by the leading and most intelligent citizens of St. Landry, members of the agricultural society of that parish, states:

1. White men have tilled the soil of this parish from the early settlement of this State, and are now tilling it without experiencing any difficulties.

2. One white man easily cultivates 20 acres of land here under the old system, and much more by the use of labor saving machinery lately introduced, and from his individual exertion he can reap a larger return than in the Western or Middle States.

3. Out-door labor to the white man is not unpleasant, and can be sustained without detriment to health. The average temperature of the winter is about 40 degrees, and that of the summer 85 degrees during the day. During the heated period we have none of the sultriness of the more Northern States at night. This has struck every one as a remarkable feature in our climate. After the heat of the day there comes a delicious feeling of pleasure produced by a cool sea breeze and the refreshing dew, which combine to cool the hot earth, tone and invigorate the system and prepare it for the coming day.

4. More than one-half of the white population of this parish are engaged strictly in agricultural pursuits, and are as robust and healthy as any similar number of farmers in higher latitudes. The population of St. Landry is 25,000, half of this number white people (1870).

5. The production of one white man's labor, who is industrious, may be thus stated: 400 bushels of corn, 200 bushels of sweet and as many of Irish potatoes, 10 barrels or 30 bushels of rice, 5 bales of cotton, with a full supply of vegetables, hay, etc.

This statement is the actual result of one man's exertions, and may be safely taken as a fair average. We appeal to the experience of five hundred white farmers in our midst for the truthfulness of every fact given, and do not fear contradiction in one single particular.

The same document, treating of the cultivation of sugar cane by white labor, says:

White men can insure to themselves a fair revenue by the cultivation of sugar cane upon a small scale. It opens the way for co-operative associations. The men of means, by putting up the machinery, can thus secure to the small farmers near him the means of saving their crops, and thus stimulate the enterprise of the people. His investment will return to him a good interest.

PROFITS OF WHITE LABOR ON SOUTHERN FARMS.

DR. TAYLOR'S STATEMENT.

The following statements were furnished us by Dr. Taylor:

Last year, (1869,) Mr. Lemoradier, assisted by his brother in-law, made 21 bales of cotton, a large crop of corn, and performed all the labor of the farm, attending to stock, gardening, etc. Mr. Dunbar rented the Eureka plantation of W. O. Denegre, Esq., of New Orleans. His four sons, the youngest sixteen, his son-in-law, Mr. Simpson, and two negro men, seven in all, made 77 bales of cotton and 1200 bushels of corn. It is worthy of remark, that none of these white men were accustomed to farm labor. Mr. Dunbar sub-rented portions of the plantation to Messrs. Humble and four Germans.

The two Messrs Humble, with their two sons, one 12 and the other 13 years of age, one negro boy and a negro man, made seventy bales of cotton, 609 bushels of corn, and cultivatad and sold 200 bushels of Irish potatoes, and cultivated three acres of cane, which was sold for $450. They hired some labor in scraping their cotton, and had some assistance in picking.

Mr. Duckworth, with one young man and two boys, made 32 bales of cotton and an abundant crop of corn. Mr. Duckworth did but little himself, so the crop was made chiefly by the young man and boys.

The Messrs. Gilette, father and son, in addition to a corn and cotton crop, cultivated some sugar cane. They made 12 bales of cotton, and 16 barrels of syrup on a small sorghum pan. The younger Mr. Gilette is a promising young minister in the Methodist Church, and attended during the year to his clerical duties.

These examples prove that white labor is eminently successful in this parish. I don't believe there was a case of sickness among these people during the planting, cultivating or harvesting season. All experience has demonstrated that, while the white laboring man in this climate enjoys equal exemption with the negro from disease, his labor from his superior intelligence and energy, is much more remunerative.

I have always found white laborers who work in the field in St. Landry the year round as healthy as field laborers in any other States—as healthy as white men who do not labor; and that this fact holds good in regard to white men who were raised in affluence and ease, and who have done their first field work since the war.

J. A. Taylor, M. D.

DR. TATMAN ON WHITE LABOR.

The following extract from a letter written by Dr. Tatman, of St. Landry parish, in 1869, and published in several papers in this State, will be read with interest by all who desire reliable information about white field labor in Louisiana :

I have been asked the question frequently by strangers if I thought white men can perform continuous field labor here; and my answer has invariably been that I know they can. Ever since I have lived here I have been seeing white men and boys working in their own fields, frequently alongside of their own slaves, cultivating the usual crops of the country; and I have never known them to be compelled to desist from such labor except from inclination.

I have known many white people who had no slaves to work for them who have worked continuously in the field, without injury to their health, raising good crops of corn, potatoes and cotton. I have not seen that white people who perform field labor are any more liable to have fever and other diseases than those who have lived in idleness. To the contrary, I have found the idle more liable to be sick than those who work out of doors, as idleness has generally engendered habits of indulgence in alcoholic stimulants that induce disease. Persons coming from the North or from Europe need not fear that they cannot work on plantations or farms here, as the summer is no hotter here than it is in the West, and the work no harder.

HEALTH OF WHITE LABORERS IN SOUTHERN LOUISIANA, IN SUGAR, COTTON AND RICE FIELDS—MECHANICS.

There is no greater error than that which generally prevails in regard to the health of out-door white laborers in Louisiana. All along the southern portion of this State, near latitude 30 degrees, between

30 and 31 degrees, and to the very edge of the sea marsh that is daily washed by the waters of the Gulf of Mexico, there are thousands of white men who work in the open air every month in the year; and they are as healthy as negro laborers, if not more so. In the saw mills and forests around Lake Charles, in the prairies and woodlands, and on all the farms of Calcasieu and Cameron parishes, on sheets of soil elevated but a few feet above tide water in and near the sea marsh, on the Island of Grand Cheniere, Pecan Island and Cheniere Au Tigre, west of Vermilion Bay, and in the midst of the mosquito kingdom, there are many white families, hundreds of them who work in the open air the year round, and have good health. In Vermilion parish the most of the farm labor has always been performed by white men, before as well as since emancipation. In Lafayette parish there has for thirty or forty years been a large number of white farmers who neither owned nor hired negroes. They did their own work. The Creole women have always been remarkable for their industry in these parishes. They formerly made, by hand labor, immense quantities of Attakapas cottonade that clothed multitudes of men.

Take the aggregate of these seven parishes, as defined by their present boundaries, Iberia, St. Martin, Lafayette, Vermilion, St. Landry, Cameron and Calcasieu, all lying on or near the Gulf coast, between 29 degrees, 40' and 31 degrees, and more than half of their field and open air labor has been performed by white men ever since the country was first settled.

IN THE SUGAR FIELDS.

A great deal of the open air labor of St. Mary was performed by white men in the early settlement of the parish. The foundation of many of the fortunes made in that parish was laid by white labor.

At the present time there are white men working in the cane fields in the lower part of the parish, near Berwick's Bay. Swedes and Danes, fresh from Europe, have worked in the sugar crops there the year round, and have had good health and have made money. White men work in the cane fields of Bayou Sale and Bayou Cypremont, near the sea marsh, and on the Teche, in the parish of St. Mary, and stand the climate well. More than three hundred white men, heads of families, work at the plow and hoe in the Cypremont and Au Large prairies, in the parish of St. Mary and Iberia, and have no more trouble about sickness than they have in the prairies of Illinois.

Twenty Danes, almost fresh from Europe, worked on the Garrett estate, now the property of the Berwick heirs, on Bayou Sale, parish of St. Mary, not fifteen miles from the salt water of the Gulf of Mexico last year. They made about ten hogsheads of sugar and 15 barrels of molasses to the hand, besides corn, potatoes and forage. They had good health, and have contracted to make another crop on the same place.

IN THE RICE FIELDS.

There are large numbers of white men, heads of families, who cultivate lands and make sugar, corn, rice, tobacco, and other crops, in the parishes of Terrebonne, Lafourche, Jefferson, St. Charles, St. James, and Plaquemines and other parishes near the Gulf coast. More than three-quarters of the Louisiana rice crops are made by white men.

WHITE AND BLACK LABOR—NEGROES AND MACHINERY.

Nearly all of the sugar houses, gin houses, dwellings and bridges of this State were made by white men. New Orleans was built principally by white men, was paved by white men. White labor is yearly increasing and colored labor is decreasing in this State. Negroes have an innate antipathy to new and improved implements and machinery on farms and plantations. They are very destructive to such machinery, and always use it with a mental protest against it. In proportion as intelligent white labor replaces negro labor will our lands be more economically tilled, and new and improved implements of machinery will be introduced. When white men take off their coats to take to business in this State, something will be done; farming will pay, and industrious and enterprising strangers, and Providence, will give us a helping hand. The best way to secure the right kind of immigration is for white men who live in the State to prove that a white laborer can stand this climate, and lay up money at farming.

WHAT $150 MAY DO.

An intelligent farmer in the pine lands of Louisiana, formerly a resident of New Orleans, has a wife and several children, lives well, has not been farming many years, informed us in a late conversation with

11

him, that he can support his family comfortably, from cash sales of produce to the amount of $150 a year; provided he makes all of his home supplies which he conveniently can make, and exchanges with merchants for goods such small articles as can be easily made, and spared with no inconvenience to the family or farm.

Why, then, cannot any industrious and economical family make a living on a Louisiana farm? The necessaries of life are few and cheap; the luxuries and superfluities of life are expensive and numberless.

WHAT A POOR MAN CAN DO IN THE SOUTH IN ONE YEAR.

Editors of our Home Journal:

Some time ago I told you I would send you a paper showing what can and has been done by a new settler in the pine woods. I find it a more difficult task than I presumed—there is so much to say to give an adequate idea of what has been accomplished, that I am afraid in this short communication, I shall fall far short of my intentions. If, however, it will be the means of stirring up an interest in this section of country, I shall be more than satisfied.

About the early part of October, 1874, Mr. ———, a ship carpenter pretty well advanced in years settled on a tract of land that he had just purchased in this parish with the intention, as he then said, of trying to make a living in the pine woods.

His land was a thicket of pine sapplings and briars. No house to live in—there was an old house on the place, and this he took down, and with the material put up a smaller one—sufficient for himself, wife and daughter—then commenced clearing his ground for a crop.

He was without team, so built himself a light cart that could be drawn by hand, wherewith, aided by his wife and daughter, he hauled logs and rails to fence in about 6 acres of land, of which however only 4 were actually cultivated, On these four acres he raised 35 bushels rice, 75 bushels sweet potatoes, 2 barrels Irish potatoes, 2 barrels cow peas, 2 bbls peanuts, 8 bbls corn in slip-shuck and 600 stalks of sugar cane, besides vegetables consumed during the summer.

The work was done with the hoe and axe—excepting only about ten days work plowing and harrowing through the whole year, digging sweet potatoes included. Being without capital, Mr, ——— had to give his individual labor to pay for mule and horse hire—in other words exchange work.

This does not look like any great thing, but when you see as I have seen a wilderness transformed to a comfortable home in so short a space of time, it makes one wonder what this country could be, if we would only take hold with a will and not only talk, but work.

We must bear in mind that Mr.——— is a mechanic (not a farmer by profession) and his family never labored in the field until this year—also that this is a first experience, and what he has done gives promise of better things in the future.

Hoping that I have not trespassed too much on your valuable space, I remain, yours, etc., C. H. T.

Tangipahoa Parish, La., Dec. 31st, 1875.

WHAT TWO MEN AND A PAIR OF MULES CAN MAKE IN LOWER LOUISIANA IN SIX MONTHS.

The following article was written by one of our most successful and reliable sugar planters, for the *Courier Journal* of Louisville, Ky. We copy it for the purpose of showing to our readers, both North and South, what can be done in Louisiana by a proper system of economy and labor:

Editor Our Home Journal, (New Orleans):

We have lately been so engrossed with our miserable political condition that we have lost sight for a time of our many natural advantages in location, soil, and climate, not only to live well and pleasantly, but to make money. My attention has lately been called to this fact, by observing the efforts of a quiet, industrious Creole neighbor, in that line. This man, with the aid of a pair of mules and a hired negro man, net over $3000 by six months work, upon about sixteen acres of land. It is true, he is living within twenty miles of New Orleans, a good market, but the same results may be obtained anywhere in lower Louisiana. As it may be of interest to many living in less favored climes to know how this is done, we will give the details. He has four acres in cabbage, four in early corn, (the ordinary white corn of Kentucky, that is in ear here in May,) four acres in onions, and four acres in Irish potatoes. The onions are planted in December, the Irish potatoes in January, and the corn and cabbage in February. All will be off the ground and sold in May, in ample time to replant the same land with a crop of the ordinary yellow corn and cow-peas, yielding from the sixteen acres not less than $200 in corn and $200 in pea-hay. The latter has not only this money value, but is invaluable as a fertilizer of the soil for the same crops the succeeding year.

Onions yield from 50 to 100 barrels per acre, and are now worth $8 per barrel, although the usual price is about $4. Irish potatoes are worth about $4 per barrel, to be sold in New Orleans or shipped to the West. They are somewhat an uncertain crop, from too much rain, but usually yield 50 or 60 barrels per acre of marketable potatoes. Cabbage are worth from $5 to $15 per 100, according to size, and the usual stand is 4,000 per acre. Green corn sells from $3 to $5 per hundred, and with a proper stand in five-foot rows, will yield from 6,000 to 7,000 ears per acre.

Now, the gross product of these crops, with an ordinary yield, and at ordinary prices, would be, say:

Onions, 240 barrels at $4	$960
Potatoes, 240 barrels at $4	960
Cabbage, 16,000 head at 7c	1,120
Green corn. 25.000 ears at 3c	750
Total	$3,790
If the producer hired the hauling to market, say two carts for 50 days at $5. to be deducted	250
Leaving the net proceeds	$3,540

The four hundred dollars' worth of corn and pea hay, being needed to feed the pair of mules, I have left out of the estimate, but should properly be included, thus swelling the amount to very near $4000. Save the gathering of the corn and pea hay in August, the two men and pair of mules would be idle, unless employed at other business, from June until November. It is usual to begin plowing, preparing. etc., for these crops, in November.

In conclusion I would say, if the producer of this crop owned the land, and wished to enhance the value of it enormously, he could plant this sixteen acres in orange trees, one hundred trees to the acre, without any detriment to the trees or crop, until the former are in bearing, which would be in seven or eight years. Well grown orange trees will yield upon an average 1000 oranges to the tree, and 2000 are not unusual, but 1000 per tree at $1 per 100 would yield the proprietor of this sixteen acres an annual income of $16,000. So it will be seen that lower Louisiana can produce other things besides sugar and rice.

Lands can be bought within twenty miles of New Orleans, often with dwellings, at from thirty to fifty dollars per acre, including woods and arable land. They are alluvial and inexhaustible. We want immigrants—intelligent white men and women.

The editor will furnish our address and post-office to any one requesting it.

KENTUCKIAN.

St. Bernard, La., April 19, 1875.

WHAT OUR MEN AND BOYS HAVE DONE, AND CAN DO.

From the Clinton (La.) Patriot Democrat.

We take pleasure in laying before our readers the following statements of the farming operations of a few of our citizens, which establish that the cultivation of cotton by white men and boys pays in this locality:

Statement of Mr. Agrippa Gayden, aged 18 years:

Paid rent for 14 acres of land at $6	$ 84 00
Paid hire for one horse	30 00
Paid feeding one horse	50 00
Paid for hired labor	50 00
Total expenses	$214 00
Proceeds of seven bales cotton	$621 39
Value of 200 bushels corn	200 00
Total	821 39
Net profits	$607 39

or over $50 per month earnings.

Statement of Mr. Alexander Norwood, aged 17 years, for 1872:

Paid rent for 30 acres land, at $10	$300 00
Hire for one mule	30 00
Feed for one mule	50 00
Use of implements	10 00
Hired two boys, aged 18 and 14 years	200 00
Paid board for two boys	150 00
Total expenses	$740 00
Net proceeds of 18 bales of cotton	1,374 86
Value of 300 barrels of corn	800 00
100 bushels potatoes	75 00
Proceeds from garden	100 00
	$1,846 86
Net profits for a boy aged 17 years	$1,106 86

Statement of William Currie, aged 18 years, and Edward Currie, aged 14 years, for 1872:

Rent for thirty acres of land at $8	$240 00
Hire for one mule	30 00
Feed for mule	50 00
Total expenses, exclusive of board	$320 00
Net proceeds of 14 bales of cotton	$1,068 65
Value of 250 barrels of corn	250 00
100 bushels of sweet potatoes	75 00
4000 pounds of fodder	40 00
100 bushels of peas	100 00
50 bushels of Irish potatoes	50 00
	$1,583 65
Net profits for the two	$1,263 65

Statement of Mr. S. D. Heap, of Darlington, La.:

On his Brookside place, three Western men, in 1870, made 18 bales of cotton, 1000 barrels good corn, and 3000 pounds of fodder.

In 1871 he hired a Swede laborer, who made 300 barrels of corn on seven acres, and four bales of cotton on three acres and sixty-six rods.

In 1872 he made 42 bales of cotton and 800 barrels of corn with six men and four women, all colored.

His son, aged 14 years, in 1872 made four bales of cotton on five acres, and 160 bushels of corn on four acres, being forty bushels of corn to the acre.

The land cost Mr. Heap about five dollars per acre in 1868. He has since sold a part of it at ten dollars per acre.

The following statement is from one of our thrifty farmers, who requests that his name be not published:

Value of land and improvements	$4,800 00
Value of 14 mules and horses at $160	2,240 00
Value of wagons and implements	600 00
Total investments	$7,640 00

REVENUES FOR 1872.

Net proceeds for 80 bales of cotton	$5,760 00
Value of 1500 bushels of corn	1,500 00
Value of 10,000 pounds of fodder	100 00
Value of peas, sweet and Irish potatoes	200 00
Gross revenue	$7,560 00

EXPENSES FOR 1872.

Estimated wear and tear	$350 00
Taxes, State and parish	180 00
Rations to 17 laborers, meal and pork	680 00
One third cotton to laborers as wages	1,920 00
One-third corn and fodder as wages	533 33
Feeding 14 horses and mules	700 00
Eight per cent. on investment, $1640 00	611 20
Total expenses	$4,974 53
Net profits	$2,585 06

In this statement is not included any profits arising from raising hogs, cattle, sale or consumption of butter, poultry, eggs or wool, all of which should have been estimated.

The net revenues amount to a little over thirty per cent. of the whole amount invested; to which, if the eight per cent. interest on investment is added, it amounts to a fraction over forty-two per cent. on investment.

The following statement is also from one of our leading farmers:

Value of land, improvements, etc.	$12,000 00
There are 500 acres in cultivation, worked by 20 men, 15 women, and 10 boys. In 1872 they made 185 bales of cotton	14,800 00
4000 bushels of corn	4,000 00
1500 bushels of sweet potatoes at 75c	1,125 00
Gross revenue made by laborers	$19,925 00
He received as rent 40 bales of cotton	3,200 00
And 650 bushels of corn	650 00
And raised besides:	
15 head of cattle at $10	150 00
20 head of hogs, 150 lbs. each, 3000 lbs., at 10c. per lb.	300 00
13 head of sheep at $2 50 per head	32 50
500 lbs. of butter at 40c. per lb	200 00
Poultry and eggs sold	75 00
Net profits	$4,632 50

which gives a rent of over nine dollars for every acre cultivated.

These statements we could multiply, but consider those above given to be amply sufficient to show what has been and can be done in East Feliciana in raising cotton. The lands which yield these revenues can be bought at ten dollars per acre. We commend these facts and figures to the Western men, and ask them can they do as well? Here we do not have to burn corn for fuel, because the cost of transportation will exceed its value. Here our people can labor the whole year in the open field. Here we do not gather forage one-half the year to feed cattle with the other half. Here we have no snow storms, no northers. Here we have a mild and beautiful climate, splendid streams of water, magnificent forests of timber, and a fertile soil, that will produce more per acre than any agricultural country in the United States. Our productions are unequaled in variety and value; and here is abundant room for as many as choose to come and make for themselves happy homes.

WHITE TENANTS—WHAT THEY DID IN PLAQUEMINES LAST YEAR.

From Our Home Journal and Rural Southland.)

MR. EDITOR—As the crops of 1874 are now harvested and marketed, it would be well that reports of the results be made public, that planters and others interested might know what has been done and what success was realized by the different kinds of labor. I am able give a reliable statement of the "white tenantry system," as practiced on Forlorn Hope," Guerrier & Auger's plantation, of the parish of Iberville, for the crop of 1874, which I here submit:

Frank Huth, ground 20 acres, made 31 hogsheads.

Joseph Brown, ground 12 acres, made 24 hogsheads.

Sost. Comeaux, ground 10 acres, made 10 hogsheads.

Numa Ramoin, ground 10 acres, made 12 hogsheads.

Rich. Thomas, ground 10 acres, made 13 hogsheads.

Of the above, Huth and Brown, employed occasionally help. Thomas was in a measure assisted by his sons, two little boys; and Comeaux and Ramoin went it "alone." Huth is a German, Comeaux and Ramoin creoles, Brown a native of Louisiana, as also Thomas, I believe.

Besides the acreage ground, a portion was put down for seed, and as much land was cultivated in corn and peas as they had in cane. In addition to the field crops raised by these men, they realized a natural increase of their stock, which was pastured free of cost on the waste land of the plantation, to say nothing of the contributions to the larder by the dairy, poultry yard and garden.

These tenants delivered their cane standing, and receive one-third of the sugar and molasses, and all of the forage crops. Let the doubting Thomases ruminate. Give us a good government, rigid police and bearable taxation. It is all that we want.

I would beg of you, Mr. Editor, to call upon the sugar planters generally to give their experience in the use of fertilizers. It would be very costly indeed if every planter had to wade through all of the manures in the market to find out which is the best. By comparing notes on this subject the expense of experimenting would be immeasurably reduced to each individual, and would redound to the benefit of all. Herein lies the difficulty with our Louisiana sugar planters. There is not enough of interchange of ideas and intercommunication generally. Planters of life-long residence in the same parish are frequently found who are unacquainted with each other. This is simply inexcusable for men who should be bound together by a community of interest. It also works sadly against them whenever the "labor and wages" question comes up. Respectfully,

E. GUERRIER.

Plaquemines P. O., Iberville Parish, La., Jan. 20, 1875.

HEALTH AND LONGEVITY IN LOUISIANA.

STATEMENT OF DR. TATMAN.

Dr. C. D. Tatman, a native of Maryland, in a letter written to a committee of inquirers in 1869, gives his views with great clearness and force in regard to the diseases of southwest Louisiana, where he practiced medicine more than thirty years. Dr. Tatman is one of the most honorable physicians in the State, and his talents and learning none who know him will question. Those who desire correct information in relation to the diseases of Louisiana, should carefully examine his letter which we publish below:

BAYOU CHICOT, Parish of St. Landry, June 5, 1869.

Messrs. Thomas H. Moore and others;

GENTLEMEN—Your letter of the 29th ult. was received yesterday, and in reply, I will say that I have been engaged in the active practice of medicine in this parish over thirty years, and that I have had good opportunities to observe all the forms and peculiarities of diseases common to this part of the country, in persons living on the open prairies, the wooded uplands, and the alluvial or low lands; and that my observations have been made on whites, negroes, mulattoes, mixed bloods and aborigines.

The most common diseases seen here are those caused by change of temperature from warm to cool, or what are called colds, producing occasionally serious diseases of respiratory organs, such as pneumonia and pleurisy. But I do not suppose such diseases occur more frequently here than they do in other warm countries, and I know they are not near so common here as they are in the colder regions of the United States.

As a general rule diseases of the respiratory organs are easily managed, and yield readily to medical treatment; but our largest bills of mortality are caused by this class of diseases, as is the case everywhere else. This class of diseases is not so dangerous here as in colder climates; and by the use of sufficient woolen clothing in cool and rainy weather can be kept off altogether.

CONSUMPTION.

Consumption seldom originates here, and those who have it in its incipient stage at the North and come here have usually got well.

INFANTILE CROUP.

Infantile croup, that terror of parents, (and physicians in other places), is almost unknown here; I have not seen a case of it in the last fifteen years.

RHEUMATISM.

Rheumatism, another disease supposed to be caused by cold, is of a rare occurrence here; and in its acute or inflammatory form, is still more rarely seen. People who spend their youth and manhood in laborious occupations need not dread to have their old age tortured by pains, as is the case in nearly all cold regions.

MALARIOUS DISEASES.

What are called malarious diseases, such as intermittent fever, remittent, or billious fever, dysentery, and diseases of the liver come next in order; but they are not so common, nor anything near so dangerous as those caused by cold. I am well aware that this assertion contradicts the commonly received opinion outside of the State; but I care nothing for that. I speak from what I have been seeing and noting for the third of a century in this parish.

INTERMITTENT FEVER.

Intermittent fever occurs more or less every summer and fall; but it is seldom of a serious character, and is always easily relieved. With the old and infirm, and with those persons whose constitutions have been broken down by organic diseases, or by dissipation, it is often attended with danger (and in their cases it usually becomes congestive), and is then more likely to prove fatal unless promptly treated.

REMITTENT OR BILIOUS FEVER.

Remittent or bilious fever occurs at the same seasons with intermittents, and is as easily relieved. A large majority of the cases of these two forms of fever will get well spontaneously with rest and care; but I am free to confess that experiment might be attended with some risk.

These two forms of fever are our peculiar fevers, and I regard them as being almost devoid of danger where commou prudence is observed. The mortality arising from them, according to my experience, has not been more than one-fourth of one per cent. And yet it is the great apprehension of falling victims to these fevers that deters strangers from settling in this State.

To the European, landing at New York—the risk to health and life would be many times less by coming here to live than by going to the Western States, where he will be shaken by ague in summer and fall, and frozen in winter.

That abominable form of intermittent called ague on the Atlantic coast and in the Western States, is almost unknown here. I do not think I have seen twenty cases of it that originated here since I have been practicing medicine. Soldiers of the Confederate States army, who served in States north of this, contracted that form of fever, and came home with it; but most of them soon recovered from it under the benign influence of our mild climate.

DYSENTERY.

Dysentery occurs occasionally during very hot and wet summers, but it is not common, nor is it usually fatal. Negroes usually have it much worse than whites, and sometimes they die from it. Whites rarely have it severely, and very few of them ever die from it.

DISEASES OF THE LIVER—MEASLES—SCARLET FEVER—SMALL POX—TYPHOID FEVER.

Diseases of the liver are of rare occurrence here, although our country has the reputation abroad of being like India in that respect

I have no hesitation in saying that no warm country where there is moisture enough to maintain a good vegetable growth is so free from liver affections as Louisiana, and particularly the parish of St. Landry. No one need fear such diseases here. I have found what are called infectious diseases, as measles, whooping cough, scarlet fever, small-pox, and typhoid fever, are generally milder in their character and more easily cured here than in colder regions. They are not more common here than in other countries—no civilized country being entirely free from them.

WHITES AS HEALTHY AS NEGROES—NORTHERN MEN AND EUROPEANS.

Whites are not more liable to disease here than negroes, but their probabilities of recovering from them are greater. This arises probably from their superior intelligence and greater prudence.

People coming from other States and from Europe are not more liable to have fevers than those born here, and when they have them their chances to get well are as good as those of the natives of the State. This assertion is another flat contradiction of the commonly received opinion outside of our State. The doctrine is taught in books and orally that all persons who settle here must undergo what is called an acclamation fever, from which the chances to recover are not great. This is a foolish error, gravely taught by sincere men, who believe it, as well as by designing men who do not care whether it is true or false.

ACCLIMATION.

What is called acclimation in persons coming from cold to warmer climates, is that the skin takes on a portion of the functions performed by the lungs in the colder regions, and if it be assisted by wearing a little extra clothing, and keeping it clean for the first year, there is no risk of the acclimation fever at all, or any other fever, unless it be an infectious one, which can be avoided here as elsewhere by keeping away from the infection.

The wearing of coarse cloth or flannel next the skin during the first summer and fall here will protect the person from all our common diseases, and the discomfort from the prickly heat is entirely got rid of, as it will not appear under such clothing unless the cold bath be imprudently used. Let the stranger use tepid baths and wear coarse cotton shirts and drawers, or flannels, and he need not fear any acclimation disease.

LONGEVITY.

As to longevity, I think people live as long here as they do anywhere. The oldest man I ever saw died in this parish, in 1849, at the age of one hundred and twelve years. He was a native of Port Tobacco, in Maryland, and had lived here eighty years. His descendants are very numerous, and one of his sons, over eighty years old, met an accidental death a few months ago.

A woman, one of my neighbors, died in 1865, aged one hundred and six. She was a native of the hill region of South Carolina, and had lived here over sixty years.

A great uncle of my wife, died in 1863, aged ninety-five years. He was a native of the upper part of South Carolina, and lived here over fifty years.

In proportion to the population, I have known more very old people here than I ever saw in other places where I have lived; and at the present time I know several that are verging on ninety.

This State has been too recently settled to have a very large population of very old people, as the first settlers and their immediate descendants had too many hardships to encounter to live to very old age, and they shared the fate of pioneers generally, and died young; but since the country has been settled, the dangers to health and life that the first settlers had to contend with have disappeared, and our population of very old people will continue to increase.

The probabilities of living to an extreme old age are greater in warm countries than in cold ones, as two-thirds of mankind die from diseases of the respiratory organs, which are much less prevalent and dangerous in warm climates than in cold ones.

I know that it is believed at the North, and in cold countries, that life is necessarily short in warm countries. I believed so myself before I had an opportunity of correcting the error by personal observation.

I know now that the average duration of human life is longer here than it is anywhere else in the United States; and the chances for persons coming from other States or countries are as good as those of the native born to live to old age.

DOTAGE—SECOND CHILDHOOD—CRONES.

In old age here there is less of what is called dotage, or second childhood, than in other places where I have known old people; and men and women retain their mental faculties generally until they die from old age. Among all the old white women, rich or poor, that I now know, or have known here, not one has ever shown a particle of that peculiar intonation of voice and conduct that entitles them to the appellation of crones, which is so common elsewhere, and particularly in Northern countries. Among negro women it is common when they grow old. This is the country to grow old gracefully in.

INDORSED BY MESSRS. HILL AND TAYLOR.

Drs. George Hill and John A. Taylor, both distinguished physicians, the former a practitioner for 34 years, and the latter residing here for nearly the same period, concur in the above statement as to the healthfulness of this country.

DR. TAYLOR'S STATEMENT.

Dr. J. A. Taylor, a native of Maryland, a planter and stock raiser, has lived in St. Landry thirty-three years—the first seven years a practicing physician. His opinions are entitled to the highest respect.

He says that fevers are milder and more manageable in St. Landry than in the Western States, or in the tide water regions of Maryland, Virginia or the Carolinas; that congestive fever is of rare occurrence. Typhus and typhoid fevers, which prevail so extensively and with so much fatality in solder climates, he says are not a disease of this latitude, and that he has never seen but two cases of typhoid pneumonia.

Inflammatory diseases, such as acute bronchitis, pneumonia, pleurisy, inflammatory rheumatism, etc., are very rare. Taking the average of mortality the year through, it would be difficult to find a healthier region than this; and those who desire length of days will be as apt to find long life here as in any part of the United States.

There have been two yellow fever epidemics in Opelousas and Washington in thirty-three years, and they might have been prevented by a strict quarantine.

INTERESTING STATEMENT OF DR. V. BOAGNI.

OPELOUSAS, La., June 22, 1869.

To Messrs. T. Mullett, Thomas H. Lewis, and others :

GENTLEMEN—I have practiced medicine in this parish every day for the last twenty years, during which time I have traveled the distance of six times the circumference of the earth, exposed by day and night to the inclemency of the seasons, and, although of delicate form —being six feet in stature, and weighing only 125 pounds—I have never had one day of sickness. My wife has always enjoyed good health, and we have eight children, all in good health, never having lost one; and if you will permit me to add to this, that my general experience of diseases here is, that they are mainly attributable to neglect of hygienic measures, and the improper administration of occult and patent medicines, you will readily infer that my experience is favorable to the climatic properties of this locality, as compared with the statistics of other parts of the United States.

Respectfully, VIECENI BOAGNI, M. D.

STATEMENT OF OLD AND RESPECTABLE CITIZENS OF ST. LANDRY.

THE WIDOW DEFRENE.

The Widow V. Defrene lives about 15 miles from Opelousas. She has arrived at the age of 118 years, 1869; she weighs less than a hundred pounds, is tall, straight, has very good eyesight, and walks briskly for her advanced age, and her mind has a fair amount of vigor.

STILL OLDER.

Joseph Cheasson, alias Joannes, died several years ago, in this parish, at the advanced age of nearly a hundred and thirty years. When he was one hundred and fifteen years old, he moved to Texas, and, after being in that State several years, he returned to St. Landry.

Mr. Thomas died in this parish several years since, aged over 100 years. Joseph Young died in this parish thirty years ago, aged about 115 years. He married at the age of 90 years, and his wife, 25 years of age, had a son whom he lived to see married. His widow still lives in St. Landry.

Mrs. Blaize died in this parish a few years since, aged more than 100 years.

Mrs. Daigle, aunt of Mr. Choteau, who has a lease of the Avery Salt Mines, 1869, died in Opelousas. She was nearly 100 years old.

Jesse Andrus, aged 90 years, and Major John Clao, aged 90, lately died in this parish.

A NUMEROUS PROGENY.

A respectable physician of Washington informed us that Madame Guillory, an old lady of St. Landry, before her death, could count up over eight hundred lineal descendants, all blood relations of hers.

A HEALTHY FAMILY.

Mr. Joseph Langley, 95 years old, lived in St. Landry forty-two years; he has twelve children—all of them are living. He had no physician in his family for twenty-five years. When over ninety years old, in aiding a deputy sheriff who wished to cross a swimming creek, throwing his feet up by the pommel of the saddle, he crossed without wetting them, while the young man who followed him wet both feet and half of his legs.

YELLOW FEVER.

It is not astonishing that those who have represented Louisiana as a land of jungles, swamps, stagnant water, malaria, terror and death, should magnify the evils and ravages of yellow fever.

New Orleans was settled in the year 1718, and the yellow fever was not known here, as an epidemic, for seventy-five years. If we are correctly informed, it first appeared in this city in 1793. It may die out entirely, as it has done in some of the Atlantic cities where it formerly raged at times, or be kept out by an efficient quarantine, which is perhaps the key to the exemption of New York, Philadelphia, Norfolk and other cities from this dreaded plague.

The yellow fever has never done much harm in the State outside of New Orleans, except in a few instances. It has never been epidemic at Franklin, on the Teche, but twice, in 1839, and in 1854. It has never been in Opelousas but twice. It has been in Washington, St. Landry parish, three times. The New Orleans steamers land there; they do not go nearer than six miles of Opelousas. We believe it has never been epidemic at Thibodaux but once; it may have visited that place twice.

12

We think the memory and knowledge of our oldest physicians will justify us in making these statements.

Yellow fever is usually carried from place to place in boxes of goods, particularly woolen goods, in trunks of clothing, in packages, in steamboat hulls, and in the hulls of vessels; but in most places the seeds of the disease are harmless. There are old and intelligent physicians in Louisiana who challenge any one to produce an instance in which any oné has ever taken the yellow fever from another person, or where a person taking the yellow fever in the city and dying at a private house in the country, has ever communicated this disease to the family, though he may have been nursed by the family, or may have died of black vomit. They say he may take the seeds of the disease in his trunk, or in his unventilated clothes; but that if he takes nothing but the clothes he has on, the family where he dies need not fear that he will communicate the disease to them.

For many years the yellow fever appears to have been localized in different parts of this city, while other parts were free from it. One year it was at, and near, the French Market, and at another time it was above Magazine Market, the street cars running through the infected districts a hundred times a day without danger or harm to those who were in the cars. Intelligent physicians could tell a thousand curious and interesting facts about this disease which would disarm it of much of its terror.

People of temperate, regular habits, in good health, are not apt to take it like those of irregular habits; and if they take the disease they stand a better chance to recover than those of irregular habits. One who is alarmed, or is attacked with some other disease; travelers who are fatigued and have been obliged to live irregularly on the cars or steamers, are more likely to take it than those whose minds and systems are quiet.

If New Orleans were well drained, and special attention were paid to sanitary measures in all parts of the city; if it had a superabundance of pure water for the poor as well as the rich, and were to keep up as vigilant a quarantine as they do in New York or Philadelphia, it is not probable that the yellow fever would trouble them more in this than in Northern cities; and New Orleans would then be unquestionably the most healthy city on this continent.

These statements may appear extravagant to those who consider New Orleans a graveyard, but we think we speak truthfully and understandingly. Our information comes from some of the most honorable and best informed physicians of this State, and it agrees with our personal observations during a period of more than thirty years.

STATEMENT OF NEW ORLEANS PHYSICIANS.

In the month of August, last, a meeting was held in New Orleans, for the purpose of imparting to the committee representing the co-operative societies of England correct information in regard to the health of New Orleans, and the State.

We copy the following from the Picayune:

Pursuant to previous call, a large number of regular practitioners of medicine in this city assembled at quarter-past 7 o'clock, P. M., yesterday, for the purpose of considering the report of the special sub-committee appointed at the meeting of the 24th inst., to furnish medical data to the English delegation now visiting this ciiy.

Dr. D. Warren Brickell was elected chairman, and Dr. Y. Lemonnier secretary.

The following is the report of the sub-committee, which was unanimously approved:

The sub-committee to whom was referred the duty of preparing a report, in behalf of the committee of physicians appointed at the meeting of July 24, for the purpose of drawing up a statement of the health of New Orleans, beg to remark that they have not had time to do justice to a subject of such importance, and are unable to offer anything satisfactory to themselves. Nevertheless, from the materials at hand, they are able to compile some figures illustrative of the subject, and can add some statements which will be borne out by the general testimony of practitioners in our midst.

RESIDENT POPULATION OF NEW ORLEANS.

The aggregate resident population of New Orleans may be set down approximately at 200,000, of whom about 150,000 belong to the white race, and the remaining 50,000 to the black and mixed races. During the winter months this number is increased by a floating population of perhaps 50,000 more, for which some allowance should be made in calculating the mortality per thousand, while, in fact, the estimate has been made on the basis of the permanent population. Of the floating winter population a considerable number may be classed as consumptive invalids, many of whom die here, thus swelling the mortality from pulmonary consumption during the winter months considerably above what it is for the remainder of the year.

MORTALITY,

WHITE AND COLORED POPULATION COMPARED.

Great allowance should be made also for disparity in mortality between the white and colored races. To illustrate this point, the following tabular statement is offered, showing the mortality per 1000, of a population of 200,000, without consideration of the large floating population during the winter months:

Table showing the Mortality per Thousand of the Population of New Orleans.

	ALL RACES.	WHITE RACE.	COLORED RACE.	AGGREGATE MORTALITY.
1869	30	25.47	41.84	6001
1870	36.95	30.68	51.20	7391
1871	30.29	25.21	42,20	6059
1872	30.61	25.80	42.86	6122
1873	37.52	31.06	53.68	7505
1874	33.99	28.66	47.90	6798
1st half of 1875	28.20	——	——	2963

Another fact demonstrated by the researches of Dr. Chaille, is that infant mortality among the colored races is nearly double what it is among the whites.

DISEASES OF COLD CLIMATES THAT ARE NOT COMMON IN NEW ORLEANS.

It should be borne in mind that a number of serious maladies are far less prevalent in New Orleans than in more northern cities. Among these may be named typhoid, or enterio fever, a disease almost unknown to this city. The whole group of eruptive fevers, of which, particularly scarlatina, is trifling, both in extent and severity; rheumatism, greatly obviated and mitigated by the mildness of our climate; true croup, a rare disease here, and diphtheria, less common than in colder climates; meningeal inflamations, only occasional and never epidemic here; choleraic diseases, less prevalent and malignant, (due in a great degree to the use of rain water for drinking and cooking purposes, and partly also to the character of our soil); dysentery, not occurring in an epidemic form, and generally quite amenable to treatment; puerperal inflammatory affections, comparatively uncommon, never epidemic, and amenable to treatment, (only 17 deaths in 1873, and 20 in 1874.)

CHARITY HOSPITAL.

Another point to be considered is the fact that the mortality at our great Charity Hospital only partially belongs to our city, as its inmates come mostly from the country parishes and from the other States of the Mississippi Valley.

MALARIAL FEVERS.

Malarial fevers have hitherto been justly accredited with a large percentage of our mortality, amounting in 1874 to no less than seven per cent. of the aggregate. It is, however, to be remarked, that these maladies are decreasing, both in severity and general prevalence, owing to improvements in drainage and the filling of building lots, and that a large portion of the cases treated here are contracted outside our city.

YELLOW FEVER.

For many years yellow fever has been the dread of strangers coming to our city, but it must be noted that there has been but one general epidemic since 1859, and that the ratio of mortality to the entire number of cases, in the epidemic of 1867, was only about 8 per cent.

DRS. TAYLOR AND TATMAN INDORSED.

In conclusion, we would state that we have not had time or material at hand to extend our investigations to the health of the State at large, outside the city. On this point reference may be made to the communications of Drs. Taylor and Tatman, in the New Orleans Picayune, of July 28 and 29, 1875. For further particulars touching the health and mortality of New Orleans, we refer to the communication furnished for our consideration by Drs. Jones and C. H. Tebault, of this city.

D. Warren Brickell, M. D.,
Chairman of Meeting.
Sam'l Logan, M. D.,
Chairman of Committee.
J. P. Davidson, M. D.,
Chairman Sub-Committees.
C. H. Tebault, M. D.,
H. H. Herrick, M. D.,
Y. R. LeMonnier, M. D.,
C. Cash. Turpin, M. D.,
Pierre Tricou, M. D.,
Sam'l Choppin, M. D.,
J. D. Bruns, M. D.,
Sub-Committee.

DEATHS IN NEW ORLEANS AND NINE OTHER CITIES COMPARED.

The following table, giving the number of deaths in each 1000 inhabitants of the cities mentioned below, during the year 1873, was also read. It was furnished to the Louisiana Equitable Life Insurance Company of New Orleans by Dr. Dowler:

New Orleans	30.6
Savannah	39.9
Vicksburg	36.5
Troy	34.6
New York	32.7
Newark	31.6
Boston	30.5
London	29.2
Brooklyn	28.1
Chicago	27.6

Dr. Dowler states in his report that if the negro mortality was eliminated from the death rate of New Orleans, the comparison would still be more favorable.

Drs. Jones and Tebault read very interesting reports in reference to our sanitary condition, which were also unanimously approved.

The English delegates propounded several questions, which being satisfactorily answered, brought the meeting to an end.

From causes obvious to all who are acquainted with the rural population of Louisiana, the death rates in the country are much smaller than in New Orleans.

HEALTH OF LOUISIANA AND MASSACHUSETTS.

We have it from the agents of two Northern life insurance companies, that insurance offices find risks more profitable in Louisiana than in Massachusetts. A prudent, temperate man, with a sound constitution, will live as long in Louisiana as in any other State of the Union.

A MODEL MISSOURI FARMER BECOMES A LOUISIANA SUGAR PLANTER.

HIS VIEWS OF LOUISIANA SOIL AND CLIMATE.

Mr. W. L. Larimore, who now owns and lives on the Mount Magnolia Plantation, near Baton Rouge, is a Kentuckian by birth. He owned and cultivated a plantation of 2,000 acres in Missouri, and took the premium at the Missouri State Fair, in 1867, for having the most highly cultivated and best improved farm in Missouri. In 1868 and 1869 he sold a large portion of these Missouri lands for $100 to $190 per acre, and moved to Louisiana. Mr. Larimore, in a letter addressed to the Baton Rouge Gazette and Comet, dated August 26th, 1873, says:

"I have been in this country over four years, and 96° is the highest the thermometer has stood in the summer. As to health, Baton Rouge is ahead of any other city in the Union, according to the census last taken.

"Here grass and vegetables grow all winter, and one need not have to work six months to save a crop, and then feed it out the next six months to the stock that raised it. Horses, cattle, mules and sheep are easily raised here in large numbers without grain, although corn can be raised in great abundance.

"I have had less doctor's bills to pay in the four years that I have resided here than I have ever had either in Kentucky or Missouri. No fatal diseases seem to prevail here like pneumonia, typhoid fever, and other such diseases. I do not say there is no sickness nor deaths here, but really there is less than in any country I have ever lived in.

The people of the North and West who desire to take a pleasure trip, should visit this part of the South in the heat of summer, where, in ten or twelve hours, they can visit the Lake shore, enjoy the sea breeze, fish, and eat oysters out of the salt water, and enjoy the best society the sun shines upon, either in Europe or America; where people meet and build temporary houses for tents, in which they camp during camp meeting, and at the same time luxuriate in the balmy atmosphere of the sea, and breathe on its healing waters.

PROFITS OF LABOR.

Mr. Larimore continues: "A live, industrious man, a true and skillful farmer, can make from one to two hundred dollars an acre from the rich sugar lands of this State. The question is asked, if lands may produce $100 per annum per acre, why can these lands be bought for from $5 to $25 an acre? This is the most alarming feature to the stranger, who thinks there must be something wrong when lands will not sell for a tenth the value of a single crop. That something is the great scarcity of enterprising, industrious, live men, who have a will to do and accomplish what they undertake, either with or without capital. I therefore urgently invite all such men to come by the fifties, or hundreds; yea, by the thousands; for there is room for each and all, and five times as many more.

LATENT WEALTH IN LOUISIANA SOIL.

"I would say," continues Mr. Larimore, "that the lands of Louisiana are capable of yielding a richer reward than the mines of California, and infinitely more certain to reward the man of industry, and that perpetually, for the more these lands are worked and cared for the more valuable they will become.

"Therefore I say to the man who has nothing to take to market but his industry, bring it here, invest it, and you will make large profits. And to those who have labor and capital, let them also come, for nowhere on this continent can such make as profitable investments as here.

LIVING PROOFS.

"I have an adjoining neighbor," says Mr. Larimore, "who, four years since, bought a plantation of about 400 acres. This man had no capital, yet he paid for his place from his first crop, and has now his place in first class order, well improved, and in a high state of cultivation, well stocked, and has money at interest. This is the kind of live men I wish to see immigrate to this country. This neighbor made twenty-four hogsheads of sugar on ten acres of land that others thought unworthy of cultivation.

THE EXALTED DESTINY OF LOUISIANA.

Mr. Larimore continues:

When a few of our Western and European intelligent, enterprising, industrious men come in and help to cultivate this soil, and their friends follow them, then all will go on harmoniously, and each will see, eye to eye, the common interests of our common country. Comparatively few see the high position this Southern country is bound soon to assume in rank with the States of the Union. Louisiana is bound to be one of the first, if not the very first State, agriculturally considered, in the Union. When we consider the vastness of her rich soil, the value of her products, the inexhaustible demand for them, and for those of her sister Southern States, what can withhold her from rising to the most enviable position?

OUR CHIEF WANTS.

The writer of the above states, with emphasis: "*All we want here is a class of live, intelligent, industrious, and enterprising men, and Louisiana will rise and prosper, North or no North, money or no money, railroad or no railroad, politics or no politics; all we have to do is to attend to our business and develop the resources of the country, and all other things will come as a neeessity when the resources of the country are fully developed.*"

SUMMER HEAT.

In the heat of summer, continues Mr. L., the thermoneter stands here about ten degrees lower than in any of the Northern and Western States. Here sunstroke is scarcely known. I have never felt the need of an umbrella to shelter me from the sun half as much as in Kentucky and Missouri. These facts I have experienced, and know.

INDORSEMENT BY THE BATON ROUGE BOARD OF TRADE.

At a meeting of the Merchants' and Manufacturers Board of

Trade, and many citizens of the city of Baton Bouge, and the parish, after listening to the letter from which we have made extracts above; they indorse Mr. Larimore in the following language :

"As evidence of our approval oi the many truths so plainly stated by our esteemed fellow-citizen, Mr. W. L. Larimore, in his letter written for general circulation, we recommend it to the careful perusal of all persons interested, both in the United States and in Europe.

ST. TAMMANY PARISH.

FACTS FOR IMMIGRANTS.

From the St. Tammany Parish, Louisiana Farmer.)

Persons intending to locate in St. Tammany parish should take the New Orleans & Mandeville Packet Company's boat, at the Lake end of the Pontchartrain Railroad—(C. M. Soria, agent, 18 and 20 Union street, New Orleans)—and make a bee line for the office of the St. Tammany Farmer, where all iuformation necessary for locating lands will be given.

ST. TAMMANY PARISH

Is located twenty-miles north of New Orleans. Covington, the county seat, is ten miles further north into the interior, and delightfully located on the Bogue Falya river, containing about seven hundred inhabitants, and is the best business point in the parish, commanding the trade for one hundred miles. The line of the New Orleans & Northeastern Railroad, already graded from Lewisburg on the Lake shore, to near Covington, will soon be built, which will add much to the trade and travel of the place.

THE SOIL

Is very productive, and admirably adapted for raising sugar cane, corn, cotton, oats, potatoes, peanuts, etc. All vegetables thrive well. Fruit in abundance is raised, and of a superior quality. The vine is indigenous, and yearly bears large harvests of delicious fruit. The orange, the apple, the pear, the peach, the quince and pecan thrive well and produce plenty of fruit.

THE CLIMATE

Is as fine as any in the world, the winters are like Indian summers, and snow rarely falls. Spring opens early in February, with blossoms on the peach and quince trees, and vegetation comes rapidly forward. The heat of summer is moderate, and the unfailing breezes of the evening refresh man and beast after the labors of the day. There is no stagnant water in St. Tammany after leaving the Lake shore, which makes it remarkably free from malaria. The atmosphere is pure and clear. The average elevation over the Lake is one hundred feet. Consumption, in its incipient stages, bronchial affections, and all throat diseases, yield readily to its healing influences; and it is a fact that yellow fever was never known in the parish, even in times of epidemic in New Orleans.

MINERAL SPRINGS

Are abundant in the neighborhood of Covington, equal in healing properties to Bladon and Virginia Springs. The Roach and Hosmer wells, in the town of Covington; Ingram's sulphur springs, on the Bogue Falaya, adjoining the town, where a large brick hotel is partially built; and the Abita, two miles distant; all of which are well known to possess medical qualities of the highest order.

WATER POWER.

We have an abundance of water power already improved, inviting the attention of manufacturers; amongst which are the old Penn place, on the Tchefuncta; the old Mortee mill, on Bogue Falaya; the Page and Prichard mill sites, on Little Bogue Falaya; and many others farther in the interior, the waters of which are equal to any in America for paper manufacturing and the bleaching of linen and cotton.

CATTLE AND SHEEP

Graze in our pine lands throughout the year, and seldom are fed in winter, thus making St. Tammany the stock-growers's home.

FIRE CLAY

Is found in abundance, as well as clays suitable for pottery ware of the very finest texture, purely white, and free from sand or grit. Persons having a practical knowledge of such business, would soon realize a fortune in manufacturing and shipping such wares to the New Orleans market.

SAND

Suitable for the manufacturing of glass, is abundant all along the banks of the Bogue Falya River, above the town of Covington. It has been practically tested by Pittsburg manufacturers, and found not wanting in properties to produce first-class glassware.

PRICES OF LAND.

Farming and stock lands can be purchased at rates varying from $1 to $5 per acre, and Government homesteads to actual settlers.

We have at present for sale some of the finest farms in the parish, at less than the cost of improvements, and on long credit to enterprising men, who will enter upon the work of farming at once, where levees are unknown and inundation impossible.

BUILDING SITES

For manufacturing purposes, can be had on the most favorable terms, and the utmost facilities offered for immigrants to this point. All that is lacking for the successful development of these rare resources, is a thousand or more live and enterprising Yankees, with a peculiar offensiveness for political pursuits. To all such we extend a hearty welcome, and guarantee protection in their respective industrial occupations. The early attention of farmers, fruit raisers and manufacturers seeking desirable locations are invited, and correspondence solicited.

AN INDIANA FARMER SETTLES IN ST. TAMMANY—HIS VIEWS OF THE PINE LANDS, AND THE PARISH.

A writer in the "Son of the Soil," describing St. Tammany parish, says:

I was raised a farmer, in Indiana, the land of hog and hominy; and after a residence in this State, winter and summer, since 1863, I can state truthfully that we have a much better agricultural State, and more healthy, than the old Hoosier land that I so fondly remember. So far as health is concerned, no country can boast of better than Louisiana, especially in the pine woods districts.

I have lived in the parish of St. Tammany for nearly five years, on a farm, and have been a close observer, and have come to the conclusion that this parish is destined to become the garden of Louisiana.

We have a doctor in this parish who has been trying for the past two years to make money enough to move to some country where his business will pay better than in this place.

The Abita Magnetic Springs, of this parish, have effected wonderful cures in many instances.

Mr. Milton Burns, eight miles north of Covington, cleared two hundred dollars an acre on his sugar crop last year. He has a scuppernong grape vine, six years old, which covers an arbor fifty feet square.

Mr. Henry Deutsch, a German, settled in this parish a few years since, on government land. He made $1200 on six acres of sugar cane last year.

Geo. Keep, senior, and Geo. Keep, junior, made $1000, clear cash from seven acres sugar cane, last year, and will do better this year.

We have three German settlements in St. Tammany, all prosperous. For small farming, fruits, chickens and vegetables, I think St. Tammany holds out the best of inducements, especially to Northern people.

For the information of your Iowa, Wisconsin, Illinois, Indiana, Missouri and Ohio readers, who want to get away from the cold winters of the North, I will state that I know of no country on the face of the earth that will equal Louisina.

LETTER FROM PLAQUEMINE.

A NEW INDUSTRY—THE PROFITS OF BEE CULTURE—A GENTLEMAN WHO GATHERS SIX BARRELS OF HONEY PER WEEK.

PLAQUEMINE, Iberville Parish, July 11th, 1875.

New Orleans Bulletin:

Mr. Editor—Another source of industry started here, and of which mention is made in your paper of the 9th, is bee culture. This will unquestionably prove a very profitable industry in our State.

Two gentlemen, Messrs. Michael Schlatre, of this place, and P. L. Viallon, of Bayou Goula, commenced with a few hives three years ago. Mr. Schlatre has managed his sugar plantation and had all the time necessary to attend to his apiary. He has now some seventy hives, and has realized very handsomely so far this season, and will continue to do so.

Mr. Viallon is a druggist, yet he has plenty of time, in his leisure moments, to attend to some thirty hives, which more than pay him for his time, besides giving him an abundance of the most delicious honey for all his own purposes. Some four or five other parties have started apiaries in this parish, and will next year be realizing finely from their work.

The most successful apiary I know of is that of Mr. Charles Parlange, on False River, in Pointe Coupee. Mr. P. is working about 140 colonies, and is getting six barrels of honey a week. This he ships to New York and sells at about one dollar per gallon.

When Mr. Parlange set about starting his apiary some four years ago, he was a promising young law student of eighteen. His friends tried to dissuade him from his "bee business," saying he would surely fail at it, and more than likely neglect his studies and fail in his effort to become a worthy member of an honorable profession. Mr. Parlange was not to be thus discouraged. His hours of recreation were given to his bees, and his legal studies pursued carefully. Last year he was admitted to the bar, and is now a "rising young lawyer," while his apiary is a success far beyond his own enthusiastic expectations. So much for energy and perseverance.

PELICAN.

NOTES ON THE RESOURCES OF LOUISIANA.

By Joseph Jones, M. D.

MINERAL RESOURCES OF LOUISIANA.

The cretacious, tertiary and post tertiary are the only formations which appear in Louisiana.

The cretacious strata appears in a very few isolated outcrops in St. Landry and Winn parishes, and has been pierced in several localities in boring artesian wells.

The tertiary forms the basis of the upland region of the State.

The post tertiary forms almost everywhere the surface, and is of the greatest practical importance to the agriculturist.

The cretacious strata probably underlies the whole State, rising nearer the surface than elsewhere in Winn parish, Chicot and Petit Anse.

Cretacious limestone of good quality for burning into lime, and of sufficient hardness to be used as a building stone, outcrops in St. Landry, about seven miles west of Chicot, and upon several points upon lower Saline Bayou.

The cretacious strata have been penetrated in boring artesian wells at Drake's salt works, on Bayou Saline, King's, in Castor, and the sulphur well in Calcasieu.

SULPHUR AND GYPSUM.

A remarkable deposit of sulphur occurs at Calcasieu. The following is a section of the well:

1. 160 feet blue clay and layers of sand.
2. 178 feet sand.
3. 10 feet clay rock (soapstone.)
4. 40 feet blue anthiconitic limestone, fissured,
5. 60 feet gray limestone.
6. 100 feet pure crystalline sulphur.
7. 137 feet gypsum, with sulphur.
8. 10 feet sulphur.
9. 540 feet gypsum, grayish hue,

The first four strata were all more or less oil bearing.

The sulphur is of unequaled thickness and purity, and the gypsum is also of superior quality.

This deposit alone is capable of supplying the entire country with sulphur and gypsum.

From the sulphur may be manufactured sulphuric acid, so important in the arts and agriculture.

SALT.

Salines occur in various portions of the State, and the rich supplies must necessarily excite the attention of capitalists

North of Red River, in Bienville and Bossier parishes, there are immense quantities of saline waters and saliferous deposits, the latter being especially found in the beds of ancient lakes. In the low flat beds of these basins which lie below the ordinary level of the country, the wells are sunk to a depth of from twelve to twenty feet, where the salt water percolates through the soil and furnishes an abundant daily supply. This is boiled in kettles, and each well furnishes from twenty to twenty-five bushels of salt per day.

In a line beginning about twenty-miles west of the mouth of th

Atchafalaya, on the coast of Belle Isle, and running nearly due east, are ranged five islands, Belle Isle, Cote Blanche, Week's Island, Petit Anse and Miller's Island.

The islands rise from the low marsh and prairies by which they are surrounded, and form mounds of various sizes. The chief of them is Petit Anse (Avery's Island), which is 185 feet above the sea tide level, and contains an immense deposit of common salt. Petit Anse is situated in Bayou Petit Anse, six miles from the north shore of Vermilion Bay, which is an arm of the Gulf. It is fifteen miles to the mouth of that bay, where there is a fine land-locked harbor of eight feet depth.

The following are the general results of my chemical analysis of Louisiana rock salt: Louisiana rock salt presents the form, appearance and optical properties of pure chloride of sodium. The large crystalline masses are so perfectly transparent, free from all extraneous matter, and uniform in their structure and density, that they would be suitable in all respects for the most delicate philosophical experiments upon the transmission of light through different media.

The sample of Louisiana salt submitted to analysis, as well as the largest masses, weighing several tons, are the purest and finest samples of rock salt that have ever come under my observation.

One hundred grains Louisiana rock salt yield upon analysis:

Chloride of sodium	99.617
Sulphate of lime	0,318
Sulphate of magnesia	0.062
Moisture (dried at 300 degrees)	0.093

The Louisiana rock salt contains less than one-half of one per cent. (0.473) of those substances which may be considered as foreign, viz: moisture and sulphates of lime and magnesia, which are found in greater or less quantities, according to their purity, in almost all samples of salt.

The absence of both chloride of calcium and chloride of magnesium is important, as these salts abstract moisture readily from the atmosphere; and when existing even to a limited extent in salt, impair more or less its value, by rendering it more hydroscopic. Meats cured with salt abounding with the chloride of calcium are more prone to absorb moisture from the atmosphere.

PETROLEUM.

The oil springs of the Louisiana Petroleum Coal Oil Company are situated in Calcasieu parish, about sixty miles from the coast. The oil spring contains large masses of asphaltum, which were formed by the oil becoming inspissated. Carburetted hydrogen gas passes out of the springs in a forcible and continuous stream, and when conducted in tubes can be employed for illuminating and heating purposes.

It is supposed that petroleum underlies this section of country, and that in the hands of experienced engineers, quantities of this valuable material will be obtained.

COAL.

Lignite deposits of various degrees of purity and value underlie nearly the whole upland country, from the Sabine to the Ouachita rivers. The coal makes an excellent fuel, and has been used in Shreveport.

PEAT.

Valuable deposits of peat are found in many places near the coast, and will, when reached by railroads, furnish large supplies of fuel.

IRON.

Iron ore of good quality is scattered in immense quantities over an extensive surface of Louisiana. South of Red River iron ore is found from Ouachita to Badian River, and from the Arkansas line it extends nearly to Red River; south of this it appears in De Soto, Natchitoches, Rapides and Sabine. Bienville parish is singularly rich in iron ore, Lime, and inexhaustible forests of pine and oak, from which the necessary flux and charcoal may be obtained, accompany the beds of iron ore.

GYPSUM.

This valuable fertilizing material is found in large quantities in the saline basins of Northern Louisiana; and the fertilizing properties of the waters of Red River have been justly attributed to the vast stores of this material washed down by its numerous tributaries.

We might also enumerate marls, and pigments, and clays, of fine quality, and the nitrate and carbonate of soda, amongst the mineral resources of Louisiana.

SUGAR AND MOLASSES.

Immense capital has been invested in the culture and manufacture of the sugar cane, and I have shown, by careful chemical analysis, that the sugar cane and molasses of Louisiana are of great purity.

The best machinery and the highest degree of skill are employed in this State in the manufacture of sugar and molasses.

The relative value of Louisiana molasses is shown in the following analysis, which I have consolidated into the following table:

Composition of 100,000 *Grains, or very nearly One Gallon of Louisiana, Cuba, and Northern Molasses, by Joseph Jones, M. D., of New Orleans, La.:*

100,000 grains contain

	Louisiana Molasses.	Cuba Molasses.	L'g Island Molasses.	Northern Molasses.	Northern Molasses.
Specific gravity—grains..	1,377	1,366	1,391	1,372	1,391
Both varieties sugar—					
Crystallizable and uncrystillizable..............	76,927	66,665	55,553	57,142	57,142
Crystallizable sugar......	59,778	43,134	58,889	28,571	26,373
Non-crystallizable sugar..	16,921	23,531	26,665	38.871	30,769
Acetates and carbonates of soda and potassium....	......	501	697	241	62
Carb. glucate and acetate of lime...............	123	905	1,113	358	468
Sulphate of lime........	204	363	370	612	801
Chlorides of sodium and potassium.............	223	1,054	1,277	788	1,012
Acetate glucate and sulphate of iron.........	......	175	288	500	490
Total saline ingredients..	550	3,000	3,750	2,500	2,500
Weight of one gallon....	96,446	95,648	97,419	96,075	07,419

The percentage of the different ingredients may readily be determined in this table simply by cutting off the three last figures.

The following conclusions have been drawn from my chemical analysis of the different varieties of molasses:

1. The Louisiana molasses is decidedly superior in appearance and taste to the other varieties of molasses offered in this market.

2. The proportion of crystallizable sugar is greatest in Louisiana molasses, whilst the proportion of uncrystallizable sugar is the least. The Louisiana molasses is, therefore, the most valuable, and the best suited to the purposes of the candy manufacturer, confectioner and baker.

3. The Louisiana molasses contains far less inorganic salts than the other varieties of molasses. Thus a gallon of Louisiana molasses does not contain over four hundred grains of salts, whilst the Northern molasses contains from twenty-five hundred to three thousand seven hundred grains. This is a point of great interest, not only because these salts interfere with the crystallization of the cane sugar, but also because they act as purgatives upon the bowels. The Louisiana molasses may be considered as almost entirely free from these impurities.

4. The Louisiana molasses is entirely free from iron salts, whilst in the samples of Northern syrups the salts of iron vary from 280 to 500 grains per gallon. These salts of iron are injurious to the health, and especially to children, and at the same time they blacken and injure the teeth.

5. In every respect the Louisiana molasses is superior to each of the other samples of molasses, and combines richness and purity of composition with an elegant appearance, pure taste and wholesome action. Louisiana molasses is fourfold more valuable than the Northern molasses.

COTTON.

The great value of the cotton crop of Louisiana is well known.

We desire simply to call the attention of the planters of the Southern and Western States to one product of the cotton plant, which, up to a comparatively recent period, was thrown away, viz: cotton seed.

Several manufactories of cotton seed oil and cotton cake have been established in New Orleans. One of the largest of these is the Louisiana Oil Company.

The works of the Louisiana Oil Company will compare favorably in their magnitude, and in their perfection, and in the power and precision of the machinery, with those of any similar manufactory in the United States.

Fifteen thousand six hundred (15,600) tons of cotton seed are annually consumed by the Louisiana Oil Company, yielding three millions six hundred and five thousand six hundred (3,605,600) gallons of oil, of superior quality; and six thousand eight hundred and ninety-nine (6899) tons of decorticated cotton seed cake. The works furnish employment to about one hundred men.

On submitting to strong pressure the oily seeds of the cotton plants, (Gossypium barbadium), a valuable and agreeable smelling and pleasant tasted oil is obtained, which in a purified state is now employed for the usual purpose in commerce and in the arts, and in pharmacy, in which other kinds of oils and fats are employed.

The production of cotton seed oil has been steadily increasing, and large exportations of the oil and cake are annually made from New Orleans to England, France and other European countries.

In Europe the cotton seed cake is regarded with high favor, both on account of high nutritive value and as a superior manure. The value of cotton seed as an efficient fertilizer has long been recognized by experienced planters in the Southern Atlantic and Gulf States.

PHYSICAL PROPERTIES AND CHEMICAL CONSTITUTION OF COTTON SEED CAKE.

The cotton seed cake and meal, prepared by the Louisiana Oil Company, is manufactured from the shelled or decorticated seed, and is far superior to that obtained from the whole seed, which contains less oil, and much greater proportion of ligneous fibre.

The decorticated cotton seed cake has a rich yellow color, is free from any strong or disagreeable smell, and has an agreeable taste; it is almost entirely free from the dark brown, comparatively valueless seed shells; and it does not contain any large amount of mucilage, nor anything that produces, on mixing with water, a volatile, pungent, or injurious essential oil. Cattle take to it at once, and rapidly improve in flesh and fat, when fed upon it, and so valuable is this article regarded by the most experienced and intelligent European farmers for the feeding of stock, that it readily commands from £6 to £8 per ton.

Owing to its superior qualities as food, it offers considerable economic advantages to the feeders of stock, in comparison not only with other varieties of cake, but also with corn and wheat.

Chemical Analysis by Joseph Jones, M. D., of Decorticated Cotton Seed Cake and Meal.

One hundred parts of decorticated cotton seed and cake contain—

Water	8.65
Oil	18.75
Albuminous compounds, (flesh forming principles), capable of yielding 6.5 per cent. of nitrogen, valuable as a fertilizer, in the soil undergoes rapid change, and yields ammonia, carbonic acid and phosphates	38.50
Sugar, starch, gum, mucilage, digestible fibre, etc., (heat producing matter) valuable as food	26.60
Mineral matters, ash composed, chiefly of phosphate of lime and potash 7.5 per cent., capable of forming blood and bones in stock, valuable as a fertilizer, composed of phosphate of lime	3.04
Phosphate of potassa	3.51
Chlorides, containing chlorine 0.18	0.30
Sulphates, containing sulphuric acid 0.19	0.43
Carbonates of lime and magnesia	0.20

One hundred parts of ash of cotton seed contain—

Phosphate of lime	40.62
Phosphate of potassa	46.87
Chlorides of sodium and potassum, containing chlorine 2.42	3.97
Sulphates of soda, potassa and magnesia, containing sulphuric acid 2.59	5.65
Carbonates of lime and magnesia	2.85

NUTRITIVE AND AGRICUATURAL VALUE OF COTTON SEED CAKE AND MEAL.

The nutritive agricultural value of decorticated cotton seed cake, will be readily seen from the following table, giving the composition per ton of 2000 pounds.

One ton of decorticated cotton seed cake contains :

Oil, pounds	375.00
Albuminous compounds (flesh forming principles) capable of yielding 130 pounds of nitrogen	770.00

Starch, sugar, gum, mucilage and digestible fibre) heat-producing elements)..	532 00
Mineral matters, composed chiefly of phosphate of lime and potash, capable of forming blood and bones, and valuable as a fertilizer, especially, to grain, 150 pounds phosphate of lime	60 80
Phosphate of potash	70 20
Chlorides	6.00
Sulphates	8.40
Carbonates of lime and magnesia...........................	4.60

Each ton of decorticated cotton seed cake will yield 375 pounds of oil, 770 pounds of albuminous compounds (flesh producing principles), 532 pounds of sugar, starch, gum, mucilage and wood fibre (heat producing agent), and 150 pounds of salts. Total nutritive elements in one ton 1827 pounds.

RELATIVE VALUE OF COTTON SEED CAKE AND MEAL, AND PRACTICAL CONCLUSIONS.

1. The proportion of oil is higher than in the best linseed cake, which rarely contains more than 10 to 13 per cent. of this ingredient which is so valuable for supplying fat to animals.

2. Decorticated cotton seed cake contains a very high, and much higher percentage of nitrogenized, flesh-producing matter than linseed cake, corn, wheat, oats, barley, beans and peas, and this renders it specially valuable as food for young stock and dairy cows. As a large proportion of the nitrogen of the food is not assimilated in the system, but passes away with the fœces and urine the excrements produced by stock fed upon cotton seed cake, will be found particularly valuable as manure.

3. The proportion of indigestible wood fibre in decorticated cotton seed cake is small, and not larger than in the best linseed cake.

4. The ash of cotton seed cake is rich in bone material; thus each ton is capable of yielding 60 pounds of phosphate of lime and 70 pounds of phosphate of potassa.

5. The physical condition of the cotton seed cake and meal is excellent and eminently suited to the digestion and nutrition of animals.

6. At this late day no planter in the Southern Atlantic and Gulf

States is ignorant of the fact that cotton seed will compare favorably in its effects with the best commercial manures.

7. That cotton seed cake and meal is a potent fertilizer is evident, not only from its acknowledged beneficial effects when applied to growing crops, but also from the following well established considerations :

Plants require a large portion of vegetable and animal matter to afford a supply of carbon in the shape of carbonic acid and of nitrogen in the form of ammonia, both of which are evolved during the decomposition of organic substances. Cotton seed cake is most valuable to grain crops, because the seed of plants is in all instances the most highly azotized portion, and all plants require a supply of nitrogen to perfect their seed, which nitrogen cotton seed cake possesses in large proportion.

Cotton seed cake is quick in its effects, because fermentation and putrifaction take place almost immediately after its application to the growing plants, and carbonic acid and ammonia are supplied to the roots of plants in large proportions.

The plant in its infancy feeds upon or is developed out of the matter of the seed; and after it has developed certain fibres it begins to take up nourishment from the soil, while the green leaf or shoot which it has sent upward extracts carbonic acid from the air; if the roots find no food near, they increase in number and length and spread over a large surface; and in poor soil from this cause plants have an immense number of fibrous roots and a poor stunted stem, the numerous roots having been formed at the expense of matters which should have assisted the growth of the stem. Cotton seed cake, from its rapid decomposition and the comparatively large amount of carbonic acid and ammonia evolved, supplies the young plants with the best liquid and gaseous food at the most critical period, and consequently enables the plants to develope stunted stems and more perfect and numerous leaves; for it appears to be well established that the sooner the plant escapes from the state of transition in which it derives its food from the seed as well as from the soil and atmosphere, the sooner will its organs for abstracting its food from the air and the soil be developed, the more vigorous will be their growth, and the more efficient their use in the process of vegetation. Cotton seed cake insures also a better supply of moisture to the plants, because it possesses the power of absorbing from twice to ten times as much atmospheric moisture as the finest soils.

Cotton seed cake and meal is greatly superior to farm yard manure in soluble organic matters, and is equal to it in phosphates; and one ton equals in agricultural value at from 12 to 18 tons of farm yard composts.

One ton of cotton seed cake contains 150 pounds of saline matters, chiefly phosphates. We have in this fact a strong argument for the use of cotton seed as a manure, in order that these salts should be returned, which are essential to the preservation of the fertility of the soil.

The great advantage resulting from the application of cotton seed as a manure for corn is referable to the high percentage of phosphoric acid and potash in corn. One thousand pounds of cotton seed are capable of yielding the necessary potash for about 7000 pounds of corn, and phosphoric acid for about 2200 pounds of the same grain.

MEDICAL FLORA OF LOUISIANA.

The following letter, written by Dr. D. L. Phares, and published in the New Orleans Picayune two years since, will be appreciated by every intelligent reader:

WOODVILLE, MISS., January 2, 1873.

D. DENNETT, ESQ.:

In a recent number of the Picayune, I read, with much satisfaction, a paper on the resources and industries of Louisiana. I beg leave to call your attention to one section of wealth and source of income not mentioned in that paper, nor yet developed as an additional and lucrative industry. Making an excursion in May, some four years ago, from Clinton, (E. F.,) to Covington and Madisonville, I was most forcibly impressed with the wonderful wealth displayed by the medical flora of that region. Looking over my collections, I find in the district between the Amite and Tangipahoa rivers, one hundred and fifty seven genera of medicinal plants, comprising nearly two hundred species. Some of these, it is true, are not worth much, but many of them are very valuable. Some are in limited quantity, but others are very abundant.

I have examined the botanical productions of that district but the once; yet I am convinced, a much larger number of species of medicinal plants can be found there than above indicated. I doubt not that many thousands of dollars could be picked up in a few months of each year by collecting and preparing for market these plants; and that it can be either as a distinct industry, or by farmers at leisure times. Other portions of Louisiana are rich in the same class of products. But my intention was simply to direct your view to this element of wealth. With highest respect,

(Signed) D. L. Phares, M. D. M. M.

MERITS OF LOUISIANA SOIL—COST OF FERTILIZERS ON NORTHERN FARMS AND GARDENS—MARKET GARDENING IN LOUISIANA—SOUTHERN ONIONS.

INNATE VALUE OF LOUISIANA SOIL.

While in some of the Northern States the farmers and market gardeners spend from fifty to four hundred dollars, and more, per acre, yearly, in manures and fertilizers, in producing tobacco, and vegetables. Louisiana has a vast surface of the richest soil on earth which needs no such expenses; and, by judicious cultivation, and a proper rotation of crops, returning to the soil fertilizers that are usually thrown away, and plowing in pea vines, these lands would never wear out. And in Louisiana we have a much greater variety of field products and fruits than they have in the North.

Four thousand pounds of sugar and two hundred gallons of molasses have been produced from a single acre of sugar land in one year, without manure, and sixty-six bushels of rice have been made from an acre without manure.

SOUTHERN ONIONS.

A Swedish farmer on the N. O., St. L. & C. Railroad made, in one year, on pine lands, ten acres of onions, without hiring any labor, and sold the crop in New Orleans for three thousand dollars.

Wm. H. Cook, Esq., near Centreville, on Bayou Teche, in the parish of St. Mary, obtained, one year, a yield of onions in his garden, that was at the rate of twelve hundred bushels to the acre.

GRASSES ADAPTED TO LOUISIANA SOIL AND CLIMATE—WINTER GRASSES—WOODLAND PASTURES, ETC., ETC.

On account of the immense importance of the grass question in Louisiana—more important, we think by far, than sugar, rice, cotton and tobacco, all combined—we give a liberal space to this subject. Dr. Phares, of Woodville, Mississippi, and Mr. Howard, of Georgia, are authority upon Southern grasses, that command respect wherever they are known. We have drawn largely from their writings, and would have drawn more heavily still, had we space, and our readers patience, to justify it.

The truths they bring forth on the grazing capacities of the South are like apples of gold, and pearls of great price.

GRASSES AND FORAGE PLANTS OF THE SOUTH.

BY C. W. HOWARD, KINGTON, GEORGIA.

This valuable little work of thirty pages is awakening much interest in Southern grasses, and shows clearly that there is more money in grasses at the South than in all other crops. He says:

"It is a significant fact that the rich lands in upper Georgia, in which a mixed husbandry prevails, have rather increased than decreased since the war in value. Let the fact be pondered that the depression in *price has occured only in lands devoted exclusively to cotton and rice culture in Georgia, both of which require a large amount of labor.*

VALUE OF GRASS AND HAY.

In the Hay, Straw and Grain Reporter, we find the following estimates:

The aggregate reported value of the farm products of the United States, including additions to stock depending on hay and grass, and betterments in 1870, was $2,447,538,658.

The Hay crop of that year was over twenty-seven million tons. This, at half the selling price in the cities, would amount to $405,000,000. So much for cured hay.

The grass crop, live stock, including horned cattle, sheep and swine to the value of $1,525,000,000 subsisted on the grass crop in 1870.

Animals slaughtered, 1870	$398,000,000
Hay crop, 1870	405,000,000
Butter, 1870	128,000,000
Milk, 1870,	25,000,000
Wool, 1870,	25,000,000
Cheese, 1870	5,000,000
	$986,000,000

The amount of credit for grass which the horses and mules, and some other animals, consumed in 1870, and for the manure of animals, valued at $1,525,000,000,) it is impossible to state. The value of the farms themselves, etc., the security for all field and garden crops, and fruits, are in a great measure dependent upon the yearly supply of manure.

The true credits to grass and hay would doubtless be more than half of all the yearly farm products of the United States, and every other civilized country, if correctly estimated.

WINTER GRAZING IN LOUISIANA.

Mr. Howard, in his Manuel, remarks, that winter grazing is the great advantage of the South. It more than compensates for the draught and continued heat of summer. It saves to a considerable extent, the cost of cutting and curing hay, and of the construction of expensive barns. The writer has sold fat Ayrshire cattle, fat enough to have been approved in Scotland, which never had tasted a mouthful of food, winter or summer, save that which they gathered for themselves. The reasonable conclusion is that the Southern climate, if we consider the whole year, is well adapted to the successful cultivation of valuable forage plants and grasses.

SUCCESS IN GRASS CULTURE.

Success in grass culture in the South is simply a question of food for the plant. Different kinds of grasses require different kinds of soil. Pipe clay lands will bring neither cotton nor corn to advantage, but will produce excellent herds grass. There are tens of thousands of acres of this pipe clay land, now utterly worthless to their owners, which would make fine herds grass meadows. Land may bring a bale of cotton to the acre, and yet be poor grass land, or it may be poor cotton land and good grass land.

We have in the South comparatively little need of artificial summer pastures. Broom sedge makes excellent spring pastures, and crab grass in the stubble gives a summer pasture which cannot be surpassed. This is a point of superiority of the South over the North. The Northern farmer has nothing to correspond to our crab grass. We are fortunately exempt in Southern pastures from perennial weeds. On the whole, the drawbacks, to successful grass culture in the South are as few, and as easily removable, as in any portion of Christendom.

LUCERNE.

Lucerne hay is extremely nutritious, and is relished by horses, cattle and sheep, so far as the observation of the writer extends, it is preferred by the domestic animals to any other kind of hay.

The product of lucerne is enormous. Five tons of excellent hay may be cut from one acre of ground planted in lucerne. It is estimated that fodder, green and dry, may be obtained from an acre of lucerne sufficient for the support of five horses during the entire year,—this includes the great bulk of green food during the spring, summer, and autumn. In latitude 32 lucerne is not green during the months of December, January, and a part of February. In the low country, along the Gulf Coast, it would probably be green all the year. In this section, (Mr. Howard writes at Kington, Georgia, latitude 32 degrees), it commences its growth during the latter part of February, and gives its first cutting early in April, even before the wild grass begins to spring. It is ready to cut fully a month in advance of red clover. The rapidity of its growth is only exceeded by asparagas. The root is perennial, lasting ten, or fifteen, or perhaps, more years. The roots become as large as small sized carrots. Five acres of lucerne on this farm was destroyed by Sherman's horses and cattle. After that the ground was left riddled with holes, giving it the appearance of a locust year. The succeeding crop of corn was very heavy. This might have been expected in view of the fact stated by by Ville, that lucerne absorbs more ammonia from the atmosphere than any other crop.

Lucerne seems to be indifferent to the texture of the soil, provided it be dry and sufficiently rich. The writer has seen it grow luxuriantly on the sands of the seaboard, and the clay of the blue limestone country. But two things are required, the soil must be rich and dry.

Great efforts have been made to introduce lucerne into the Northern States and England. The soil and climate of England is not suited to it, and the Northern States are too cold.

Lucerne is a child of the sun. It is a plant of a warm climate. It grows as well in the Southern States as in France or Italy. It is comparatively insensible to drought. While we have a plant that yields hay of a *better quality, and double in quantity, as compared with any grass grown at the North, our railroads are groaning under the weight of Northern hay.*

THE FIELD PEA.

The amount of valuable winter forage that can be saved from an acre of the field pea is very great. It improves the land, and can be raised among corn without injuring the growing crop. The value of this crop, and its effects as a fertilizer, can hardly be overestimated.

BERMUDA GRASS.

I think it very doubtful whether there is an acre of land in the South, thoroughly set in Bermuda grass, (if proper use be made of it) that is not worth more than any other crop that can be grown on it. If I am right in this broad opinion, our efforts should be to propagate it. I am planting it every year on such land as does not pay for cultivation, and how much such land is there throughout the South ?

I cannot better illustrate the grazing value of Bermuda grass, says Mr. Howard, than by an instance in my own experience. Nearly thirty years ago I bought an old plantation near my place in Hancock county, Ga. It was bought at a low price on account of its being *infested* in places with Bermuda grass. I permitted a man to use thirty acres of it which were set in Bermuda grass. He had at the time a cow, a calf, sow and pigs, and a breed mare. He cultivated a little crop of corn, but never enough to feed his family. His stock lived upon that thirty acres of Bermuda grass, except for a short time during the winter, when they had access to other parts of the plantation. He remained upon his place for five or six years. At the end of that time he had twenty-five head of cattle, seventy-five hogs, and five horses. I offered him for his increase $1,000, which he refused. So much for the grazing value of Bermuda grass.

I cannot give a better illustration of the manurial value of this grass than by reference to the crops made on this thirty acres of land after the man referred to had left the place.

First crop, cotton; half a stand, on account of the mass of undecomposed sod, eighteen hundred pounds seed cotton per acre, 600 pounds of lint,

Second year, two thousand eight hundred pounds of seed cotton per acre.

Third year, sixty-five bushels corn per acre, manured with cotton seed.

Fourth crop, forty-two bushels wheat per acre.

The average product of this land without the grass would have been not more than one hundred pounds of seed cotton, fifteen bushels of corn, or ten bushels of wheat. I know of no crop that will improve land more, and certainly none that will give so large an increase with so little labor.

A gentleman in this county informed me a few days since that he had just cut from one acre of Bermuda grass eight two horse wagon loads of excellent hay.

The Bermuda and crab grass are at home in the South, clear down to the Gulf shore. They not only live, but live in spite of neglect; and, when petted and encouraged, they make such grateful returns as astonish their benefactor. I have known $114 worth of Bermuda grass sold from seven-eighths of an acre in one season.

DR. ST. JULIEN RAVENEL'S OPINION.

Dr. St. Julien Ravenel, an accurate man of science, makes the following remarkable statement in regard to Bermuda grass. The experiment was conducted near Charleston, S. C., in latitude 31°.

One-eighth of an acre of ordinary land covered with Bermuda grass was plowed in March (1874), harrowed, rolled smooth enough for the mowing machine, and fifty pounds of ammoniated super-phosphate of lime applied.

Four cuttings were obtained which yielded two thousand eight hundred and eighty pounds of hay. The hay was nicely cured, and was preferred by horses and cattle to hay brought from the North.

Bermuda grass and sheep may restore the poor, gullied old farms of the South, worn out and abandoned by cotton plantations in large portions of Georgia, must become a sheep walk before it can be restored to fertility, and the land owners become independent of the negro. Besides all the clover and meadow, of which there is a vast amount, forty-eight per cent. of the whole of the valuable soil of England is in permanent pasture.

NATURAL RELIANCES FOR STOCK IN THE SOUTH.

Without reference to the artificial or cultivated grasses, we have the following natural reliances for live stock food during all seasons of the year: Terrell grass and wild cane for winter pasture; gama, crab, and crow-foot grasses for hay; Bermuda, the sedge, and other common grasses in great numbers, including crab-grass and crow-foot for spring, summer and fall grazing.

With these reliances the poor man can begin at once to raise sheep and cattle before he can make his land rich, or purchase grass seed and cultivate improved grasses. As his means improve he may add such of the artificial grasses as may be adapted to his particular soil and locality.

When we have thus availed ourselves of these natural resources, and have added the appliances of modern agricultural science, the *South will have attained a prosperity which will make her a marvel among the people of the earth.*

WINTER GRASSES.

One of the most marked and singular advantages of the South is its ability to grow grasses that may be pastured in the winter. It is a blessing of climate which we have not yet appreciated. The raising a full supply of horses, mules, cattle, sheep and hogs for our own consumption is an absolute essential of skilled agriculture.

THE MEADOW OAT GRASS.

This grass deserves to be placed at the head of the winter grasses for the South. It has the double advantage of being a good hay as well as winter pasture grass. It does not answer well on moist land. Rich upland is the proper soil for it. On such land it will grow from five to seven feet tall, completely hiding a man walking in it. It will grow on more sandy land than most of the artificial grasses. The yield of hay per acre is large, and the quality excellent. It matures rapidly Seed sown in the spring will produce seed in the fall; the seed is ripe when the stalk is green. This is a great advantage in being able to save the seed and hay from the same crop. The amount of green food yielded by this grass during the winter is greater than from any other grass.

In any movement towards an improved agriculture in the South the first step should be the laying down of rich winter pastures.

WOODS PASTURES IN THE SOUTH.

Most of the woodland in the South is, to a certain extent, dead capital. Woods pastures, which correspond to the English parks, would, in many localities, be found profitable. The parks of England are one of its chief ornaments; an ornament which is also an utility. Fine oaks, green grass, running water, bleating sheep and lowing cattle, form a landscape which the painter attempts in vain adequately to depict. There is no reason why a large portion of the neglected woodland of the South may not be made to add to our wealth, while it at the same time fills the eye with scenes of beauty.

DR. D. L. PHARES ON SOUTHERN GRASSES.

Dr. D. L. Phares, of Wilkinson County, Miss., situated in the S. W. corner of the State, and bordering on Louisiana, has published a series of articles in that sterling agricultural journal, the *Farmer's Vindicator*, on Grasses adapted to Southern Soil, that are invaluable.

Dr. Phares is a Patron of Husbandry, a member of Wilkinson Grange, Woodville, Miss., is eminently practical, is one of the most scientific men in the South, a learned botanist; and a man of sterling integrity. What he claims to know, in scientific matters, we never have the temerity to question.

The following is the substance of one of his addresses to Wilkinson Grange:

I propose, in my remarks, to give you something that is practical and affecting your own individual interests, and those of our State in the immediate future.

The grass crop of the United States is the most important and valuable of all her crops, exceeding, by far, the value of the cotton crop.

I will exhibit to you samples of grasses, and while you examine, I will talk to you about them, that you may the better understand their character and qualities,—their culture and comparative values.

The first group I open comprises perennials, or permanent grasses. And the first specimen I hand you is

DACTYLIS GLONERATA.

Orchard grass, *cock's-foot* grass. This foreign grass has long been cultivated in Europe and America, both for pasture and hay. I place it first among the grasses, because it is hardy, furnishes considerable grazing during the winter in our climate, and abundant in spring and summer; five crops of hay, and will grow almost anywhere, whether the ground is poor or rich.

It is true, that as the soil is poorer, the shorter and less valuable the crop. When once set, this grass remains as long as you choose to keep it; and it is improved by mowing and grazing, provided it is fed a little once in three years,—oftener if you choose,—with suitable manures. *It grows and thrives here as well as anywhere in the world.* I have had it mowed, and the leaves, as cut off, measured two feet to four feet nine inches in length and the seed, stalks, or culms, from two to five feet high. I mow it two or three times during the spring and summer. The time of mowing depends upon the season and the grazing. Last year we made the first mowing pretty early in April; this year in May. I keep calves, horses, mules, and at times other stock on it, from 1st of September till March or April. I have not made any accurate trials to ascertain the quantity of hay per acre it will produce, but I get, I suppose, from two to three tons.

Rev. Dr. Watkins. May I interrupt you, to make an inquiry?

Dr. Phares. Certainly. I shall be glad to answer any inquiry from you, or any other member of the grange. I hope no one will hesitate to ask any question that will enable me to give any information in my possession touching the matters under discussion.

Dr. Watkins. Will the orchard grass grow well in the South under the shade of trees?

Dr. Phares. Yes; I have had it for years growing on wooded lands cleared of underbrush,—lands with a considerable variety of trees, holly, buckwood, mulberry, walnut, magnolia, elms, tulip, beech, several species of oak, and other species of forest trees. *It grows in such situations, too, as readily, and almost as luxuriantly, as in the open fields. It grows finely even among beech roots, up to the very trunks of the trees.*

14

I sowed one lot of two acres with *red clover* five years ago, intending the clover to help out the first year's crop, and improve the land, and gradually give place to the grass for permanent pasture; but the clover has *continued* so *luxuriant*, and yielded such *large crops* (a very heavy crop just taken off) that I have pastured scarcely at all from March to September. This is the lot so much admired by many present.

It may be sown from September 1st till March 20th (perhaps later) with success. The best time to sow will depend upon things which I will explain further on.

AGROSTIS VULGARIS.

This is the *Red Top Grass*, or *Herds Grass* of Pennsylvania and the Southern States, and the *Bent Grass* of England. It grows two or three feet high. but I have mowed some even four feet high. It makes good hay in the South, and good pasture, on lands moderately moist. It grows well in marshes, and is not injured by overflow, even though somewhat prolonged. It may be sown the same time as orchard grass, two or three bushels to the acre.

POA PRATENSIS.

This is also called *Smooth Meadow Grass*, *Spear Grass*; in Kentucky, *Blue Grass*. The first year after sowing, this grass is so small that some persons have given it up as a failure, and have plowed it up. The second year there is some growth, but this grass does not attain perfection until the third year. It grows as well here, and I think better, and during the first and second year makes a much better show, than in the far famed Blue Grass region of Kentucky, or anywhere else that I have seen; I would say that it should give from three to four tons of hay to the acre.

CYNODON DACTYLON.

This is the *Durra*, or *sacred* grass of the Hindoos. Sir William Jones, in the fourth volume of his Asiatic Researches, praises its beauty, and its sweet and nutritious qualities for cattle. For twenty or thirty years, on the common where this grass grows on the Bayou Sara road, three miles west of Woodville, hundreds of herbiverous animals have feasted, summer and winter. Mr Elliott, in his Botany of South Carolina, says: "the cultivation of this grass on the poor and extensive sandhills of our middle country would probably convert them into sheep pastures of great value: but it grows in every soil; no grass, in close rich land, is more formidable to the cultivator; it must therefore be introduced with caution." But I do not regard it as very difficult to subdue, or even to destroy.

For years this grass has furnished our live stock with choice grazing, as well as a fine quantity and quality of hay. In and around Jackson, Miss., hundreds of live stock of all kinds are kept fat by this grass; and the mowing machines may be seen on the Fair Grounds, and other places in the vicinity, making large profits for the owners by turning this grass into hay. Mr. Ravenel, South Carolina's distinguished botanist, has taken in one season, at the rate of five tons per acre of fine well cured Durra hay from the Sacred Grass.

IRIPSACUM DACTYLOIDES.

This is the *Sesame grass*, or the *Gama grass*. It is a native of our Middle or Southern States. I have seen the leaves of this grass seven feet long, and the culms ten and a half feet. All kinds of live stock eat it greedily; and if allowed free access to it, will soon destroy it: and this is about the only way to destroy it, except by the immense labor of digging and hauling is away. It may be mowed five or six times a year, and produce an immense amount of coarse, but sweet and nutritious hay. It grows well in either marsh, or dry land.

PHELEUM PRATENSE.

This is *Cat's-tail* grass, the *Herd's* grass of New York and New England, but *Timothy* of Pennsylvania. I have spots of this grass growing among the clover for some years, and fully equal to that grown in any other part of America. It will not bear pasturing well in the summer in this vicinity.

HOLCUS LANATUS.

Velvet grass, *Feather grass*, *White Timothy* of the old world. Some call it *Muskeet Grass*. It is doubtless a valuable grass for our use. I am unable to say what amount of hay it would yield to the acre.

LOLIUM PERENNE.

Rag grass, corrupted usually into *Rye* grass. This grass, from Europe, is naturalized in some of our Northern States, and makes a good pasture. A variety called *Italian Rye grass*; has been highly extolled in Europe. I have had this grass three years. For the immense number of radical leaves, no other grass, except the Kentucky Blue grass, can be compared with it. I never saw any other grass cover the ground so deeply and so densely. Knowing nothing of its management, I sowed it late; but it grew astonishingly. The season was very wet: I saved no seed. I would like much to see this fine grass tried, and prove a success.

The *Lolium termulentem* has of late years made its appearance in some of the grain fields of the South. It is probably the only one of all the large order of *Granineoe* that possesses noxious properties, as its name implies. It is said to affect cattle injuriously, and the grain ground with wheat, as sometimes occurs, and made into bread, produces giddiness and other unpleasant effects on man. It is the *Zizania* of Matthew xiii, 24-3, 36-8, transferred into the Latin Vulgate, the Italian, Spanish, and some other versions, "*Ivrie*" in French, and Darnel in most English versions; but, unfortunately, "*tares*" in King James' version. The tare, vetch, or any other bean, could be recognized as soon as it sprouted as not akin to wheat, and hence the force of the parable.

MEDICAGO SATIVA.

Medick Lucerne, Trefoil, French Lucerne. This is a valuable plant. Its Spanish name is *Alfalfa.* It is ready for use early. Last year I had it two feet high by the middle of February; this year the same height a month later. It grows best in ground that is rich, and dry, and mellow. It makes good hay which is relished by stock. It is better, however, for soiling, or feeding green, or rather wilted. The amount of rich forage it may produce is probably greater than from any other plant. It is very rich in milk and butter principles, and specially suited for feeding milch cows. Stock must not have access to growing lucerne, for by eating out the crown the plants are killed.

Some plots of it are now in fine condition that are known to have been growing for over thirty-five years, without any marks of decay. Where the subsoil can be penetrated, and is not too moist, lucerne sends its roots down ten, or even fifteen feet deep. Hence it is less affected by drought than any other plant.

M. LUPULINA.

Nonesuch, or *Black Medick*. This has been naturalized with us, but is of too little value to require special notice.

M. MACULATA.

Spotted Medick. This is a valuable plant. It was brought from Chili to California, and thence to the United States under the name o *California Clover*, or *Yellow Clover*. Many mistook it for *Lucerne*, and still so call it. This has only *two* or *three yellow* blossoms in each cluster, while lucerne has many blue blossoms in an elongated head.

I have grown this plant about twenty-five years. It furnishes good grazing from February till April, or May, a small lot of ground feeding a large number of cattle, sheep, etc. Cattle do not like to eat it at first, but it is easy to teach them, and they acquire a great fondness for it. But all the grass eating animals, including geese, know and eat lucerne greedily at first sight. Horses that refuse the Spotted Medick when green, eat it readily when wilted or dried. The last lot I sowed was in 1859, or 1860. Every year many persons passing the public roads near this lot stop and admire the luxuriant growth. For a number of years my live stock had access to it from December to March, or later, with much profit. On removing them, it shot up and spread out rapidly in April and May, in the latter month maturing an immense quantity of seed, and then dying. In June the crab grass (*Punicum Sauguinale*) sprang up on the same ground, and in August this grass, while in bloom, was mowed. In October I had a second lighter mowing. In a few weeks the Medick would be up and in full possession of the ground till the next June.

Thus for years I had the latter for grazing in winter and spring, and in August and October took off two and a half or three tons of hay per acre. The hay is better than you ever get from the West.

After a luxuriant crop of Medick, the ground is very loose and in a condition to produce a good crop of anything. One may cultivate the land every year, and make better crops of corn and cotton than on ground not occupied by the Medick, and still have the benefit of the latter for winter and early spring grazing.

MELILOTUS.

Sweet Clover. We have two species of this, *M. Officinalis*, and

M. Albo; the latter being *white Melilot, Tree Clover, Bokara Clover.* Neither of these is of agricultural value; but both interesting in the flower garden for comeliness and fragrance.

TREFOLIUM PRATENSE.

Red Clover. In this county, and in a large portion of this State, this plant grows as *promptly, and as luxuriantly, and yields as heavy crops of forage, as in any other portion of America.* In truth, from a comparison of the clover crops of Mississippi with those I have seen and heard reports of in all the States further North, it is evident that in this State there is much more certainty and less difficulty in obtaining a good catch and in maintaining a good stand; and in consequence of this, in connection with climatic influences, a large yield is annually realized in this latitude than in colder and higher latitudes, and for a greater number of years.

True, this is putting it pretty strong; but not a whit stronger than I believe the facts warrant. I have secured a good stand from seed sown in all the months from September to February, and even to the middle of March.

That which you have seen growing on the east side of the telegraph line, which so many of you have stopped to admire, and which Northern and Western men have so often gone in to examine closely, and handled, to be certain that their eyes had not deceived them, was planted five years ago; yet the hands declare the crop of this year as heavy or heavier than that of any former year. I have made on that ground at the rate of *nine thousand* pounds of good clover hay to the acre. But others, in various and widely distant parts of the State have done equally well; and some much better, I am happy to acknowledge.

The field on which this clover grew is old. The subsoil is red clay, many feet deep. It was worn out and abandoned in 1833. Red clover improves the land much more than the spotted Medick, both as a fertilizer and an ameliorator.

THE TRIFOLIUM REPHAS.

White clover grows spontaneously and luxuriantly on all deep, red clay lands, and furnishes excellent grazing from February on for some months, and in many localities in our country, it is large enough to mow. Later in the year, like the second growth of red clover, it will salivate horses and mules, and produce cholic in cows. It is one of the best honey plants, and this year it is furnishing an astonishing quantity of honey.

I had prepared four other kinds of clover to present to you but time will not permit. They are however of but little value.

I intended to speak of the *Lespidezas*, of which we have growing around us five species and several varieties, besides the notorious *Le Striata* or so-called *Japan Clover*. None of these are in my estimation valuable. I intended also to speak of the *Desmodimus*, of which we have some twenty species, some probably worthy of attention and culture.

CONCLUSION.

It is a general rule in Europe, America, and other regions, that the more extensively the grasses are cultivated the more the value of the land is enhanced; grass being the best paying crop even on lands worth from $100 to $500 an acre. The lands in this part of Mississippi, are worth about one-fifth what they were six years ago. If three-fourths of our lands were laid down in suitable grasses, I have no doubt the value of our farms would be enhanced *several thousand fold.*

HOLCUS.

The late Master of the State Grange, of Louisiana, Mr. H. W. L. Lewis, cultivated this winter grass for many years without knowing its name or history. He esteems it highly. He sent specimens to Dr. Phares, who pronounced it the *Holcas*, a native of Britain. He says the French call it *Hougue, Hougue Laineuse, Foin de Mouton*, the Dutch *Zorghzaad;* German, *Das Darrgras, Wollinges Honig-gras;* Swede, *Myskgras.*

Dr. Phares says it grows on almost any land, however poor or rich the soil. It grows finely on heavy loams, but attains greatest luxuriance on light, moist, peaty soils. It is well adapted to much of the lands of Mississippi, and a still larger portion of those of Louisiana.

Its nutritive properties consists wholly in *mucilage and sugar*, while animals, relish more the grasses whose nutritive matter is partly *sub-acid* and *saline.* That it is not from deficiency of nutritive matter, but rather excess, will be evident on comparing this with other grasses. Let us take the orchard grass in bloom.

One acre of orchard grass yields green grass, 27,905 pounds, which, dried, gives 11,859 pounds, containing 1089 pounds of nutritive matter.

An acre of Holcus, same kind of land, gives green grass 19,057 pounds; dried, 6,193 pounds; nutritive matter, 1,191 pounds. As this grass in its green state contains less water than others, it yields an immensely large percentage of nutritive matter than orchard and others, though not in so desirable a proportion and condition to suit the taste of animals. It ought to be specially valuable for milch cows and sheep, and for lean horses that it is desirable to fatten. Fed green it should be better for work stock than any other green grass.

Holcus Lanatus, known as the velvet grass, *feather grass, white timothy* of the old world, and the velvet *Mesquite* of Texas.

PINE HILL LANDS.

PROFESSOR HILGARD'S VIEWS AND ANALYSIS.

As the pine lands of Louisiana are its poorest lands, and as the State has about seven million acres of these lands, it is proper that we give the reader as far as possible, of their character and qualities.

Professor Eug. W. Hilgard, State Geologist of Mississippi, a gentlemen of rare scientific attainments and great integrity of character, in his ' Report on the Geology and Agriculture of the State of Mississippi," gives the analysis of the pine land surface and subsoil in counties bordering on Louisiana, and these analyses will give a pretty correct idea of the character of the pine lands of the two States.

PINE LANDS OF PIKE COUNTY.

Depth, nine inches.

Vegetation, long leaf pine; Post, Spanish (Red,) and (true) red oaks.

Somewhat ashy, color, yellowish bluff.

The soil saturated with moisture at 69.1 deg. fahr., lost 4.106 per cent. of water at 400 degrees.

The subsoil nine to twenty inches.

An orange-yellow, rather sandy loam.

The subsoil, saturated with moisture at 69.1 deg. fahr., lost

10.0 per cent. of water at 400 deg., dried at temperature, soil and subsoil consisted of—

	Soil.	Subsoil,
Insoluble matter (chiefly fine matter)	89.801	77.931
Potash	0.218	0.266
Soda	0.076	0.072
Lime	0.034	0.152
Magnesia	0.806	0.352
Brown oxide of Maugauere	0.072	0.091
Peroxide of iron	2.402	5.456
Alumina	3.783	11.870
Phosphoric acid	0.036	0.043
Sulphuric acid	0.038	0.035
Organic matter and water	3.446	3.261

Prof. Hilgard remarks. These analyses show, 1. That both the soil and subsoil are considerably below the avergge of native fertility, (*i. e.,*) the absolute amount of nutritive ingredients contained in them. 2. That there is but a small difference in this respect between the soil and subsoil; being nevertheless decidedly in favor of the subsoil, especially with regard to lime, in which the surface soil is usually poor and also potash. There is, however, one important difference in the retentiveness of the two materials, the surface soil being defective in this particular, and the subsoil the property in a degree somewhat unusual in materials of equal lightness.

BETTER PINELAND SOIL—SMITH COUNTY.

The following analysis of pineland soil and subsoil in Smith County shows the ingredients of the better qualities of these lands: depth, 5 inches,—subsoil, 11 to 18 inches.

	Soil.	*Subsoil.*
Insoluble matter (chiefly sand)	93.257	83.030
Potash	0.259	0.485
Soda	0.065	0.061
Lime	0.129	0.073
Magnesia	0.180	0.519
Brown Oxide of Manganese	0.146	1.153
Peroxide of Iron	1.251	4.145
Alumina	2.356	8.871
Phosphoric Acid	0.030	0.022
Sulphuric Acid	0.024	0.021
Organic Matter and Water	2.380	3.117

"These analyses," says Professor H., "show very important differences to exist between the surface soil and the underlying loam, the latter not only being calculated to improve the former in its physical properties. as shown by the large amount of moisture which it absorbs, but much richer in several of the important nutritive elements; so that if at the surface, it would constitute a soil of *average* fertility. Here, also, deep plowing is indicated as the first step towards the improvement of these lands; it will increase not only the retentiveness, but also the native fertility of the soil by mixing with it a *more fertile subsoil.*"

The loam stratum of these pinelands being rarely more than 3 to 4 feet thick, and then underlaid by loose sands, the land may in general be considered as being *naturally underdrained.*"

ANALYSIS OF PINE STRAW,

By Professor E. W. Hilgard.

Insoluble matter	65.312
Potash	5.530
Soda	0.416
Lime,	13.860
Magnesia	5.208
Brown Oxide Manganese	1.681
Peroxide of iron	0.041
Alumina	4.539
Phosphoric Acid	1.154
Sulphuric Acid	0.894
Garbonic Acid	1.479
Organic matter and water	——

"This analysis shows," says Prof. H., the pine straw to contain notable qualities of all the mineral ingredients *required by useful crops,* there being a remarkable deficiency only in soda. The conclusion is therefore inevitable that by means of pine straw properly applied we can replace the drain on the soil caused by crops. The potash and phosphoric acid contained in 400 pounds of cotton lint will be returned to the soil in about 1400 pounds of pine straw, provided the stock and seed be returned to the soil with the addition of a little common salt."

PROFESSOR S. W. JOHNSON'S VIEWS.

Prof. S. W. Johnson, one of the ablest and most practical agricultural chemists on this continent, in his admirable lecture before the *Connecticut Board of Agriculture*, says that light sandy lands need leaf mould, humus, vegetable matter, to make them fertile. By the chemical action of the roots ef growing crops, sand itself is dissolved, and the potash contained in each grain so dissolved is set free to nourish vegetation. "In addition to that," says the Professor, "the vegetable matter, the humus, is itself retentive, not only of moisture, but of ammonia, and is of itself a powerful *solvent* of the *rock*, and an energetic *loam-maker*, as the rich surface of forests and prairie testify, and as the direct observations and experiments of the agricultural chemists have proved and made plain in detail."

Common glass is made of sand melted with potash, soda, and lime. "The very germination of a seed," says Dr. Riggs, in addressing the Connecticut Board of Agriculture, "creates a chemical action. The growth of a plant will even decompose a glass vessel in which it is growing. Even a flower pot is decomposed by the plants growing in it."

This may appear to be a departure from my subject, but we offer it as a key to success in pineland farming in Louisiana. These lands have usually a superior subsoil,—they are underdrained by nature, and they have *all* of the elements of fertility in the surface soil needed to secure fine crops *except leaf mould*, or *humus*, or *geine*, as it is variously ealled.

One of the most intelligent and enthusiastic pineland farmers we have met in our travels; a physician and farmer; a Georgian by birth, stated emphatically that he is confident that a thorough farmer can make a good living for himself and family on ten acres of pineland, provided they have a good subsoil, and that when these lands are made rich, as they can be by pine straw, muck, manure, green crops plowed n, etc., they are the most pleasant lands to cultivate, and the best lands in the State.

This gentleman had oats growing nearly six feet high, at the rate of more than sixty bushels to the acre, pear, apple and peach trees loaded with fruit, a garden filled with choice vegetables of rank growth, all on pinelands originally very poor.

GUINEA GRASS AND PINE LANDS.

The editor of the Florida Agricuiturist, writes of guinea grass (*panicum jumentorum*), as follows:

Guinea grass was accidentally introduced into Jamaica, from Africa, about one hundred and fifty years ago. It has since spread over the whole island, and the neighboring West India Islands. In Jamaica large tracts of country are kept planted with it, and the finest horses and cattle are reared upon it. Some fields have been planted over one hundred years to this grass without being renewed or manured. The droppings of the cattle, and the quantity of grass trampled into the ground are sufficient manure. When the fields are kept for cutting purposes they are manured once a year. In some parts of the island the wood has only to be cut down and burned off, when the field springs up in fine guinea grass.

GUINEA GRASS IN FLORIDA.

There are three kinds of guinea grass: the hardest kind, (the St. Mary's grass), the edition above mentioned introduced into Florida in 1872. He distributed plants, and all who have engaged in their culture have succeeded. The editor says: "It is now certain that only enterprise is needed to place Florida among the attractive grazing States in the Union. The success of guinea grass in Florida, I consider established beyond a doubt. The land used in my experiment was the poorest worn out pine land, too poor even to grow sweet potatoes on. I bad grass eight feet high in some places, and cut some of it three times during the season. Every horse to which it was offered ate it at once. I made hay with a small portion, and found it to succeed admirably, and a crop of twelve tons a year can be had from one acre of good land adapted to this grass.

The hardy kinds (the St. Mary's) is suited to horned cattle; the other other kinds to horses and mules. It is propagated from either seeds or roots."

Later, the editor writes: "The grass has been tried in almost all parts of Florida with success. The birds will soon scatter the seeds all over Florida, and we will then have a country for grazing purposes superior to any in the world.

GUINEA GRASS IN LOUISIANA.

Guinea grass ha been planted in Louisiana, but we doubt if the kind suited to horned cattle (the St. Mary's) has reached this State. Louisiana ought to be even a better state for guinea grass and grazing than Florida. She has the pine lands, and every other kind of lands which this grass requires, in millions of acres. And then we have millions of acres adapted to Bermuda and other valuable grasses.

WHAT PINE LAND FARMERS MAY DO.

What pine land farmers have done, and what the best class of them are doing, multitudes of others may do.

We know of pine land farmers who make 500 pounds of lint cotton to the acre, 30 to 40 bushels of corn, 40 bushels of red rust proof oats, 2 hogsheads of cane sugar, and 3 barrels of molasses from an acre, 15 bushels of highland rice, 150 bushels of sweet potatoes, 150 bushels of Irish potatoes; one who made three thousand dollars worth of onions from ten acres; many who get an astonishing yield of grapes of various kinds to the acre; many who get splendid yields of pears, peaches, strawberries and even apples, and other fruits; many who make plenty of pork to sell, and have money at interest.

HEDGING IN YELLOW FEVER.

SUCCESSFUL EXPERIMENTS BY THE BOARD OF HEALTH.

The Board of Health has been experimenting, since 1869, in the city of New Orleans, with carbolic acid, and some other disinfectants, as a means of hedging in and smothering yellow fever in any part of the city, as new cases occur in the early part of the season.

Mr. G. W. R. Bayley, a member of the Board of Health, has lately published in the Picayune the following interesting facts gathered from the Annual Report of the Board of Health for 1873.

When a case of yellow fever occurs, and it commences spreading, it travels about forty feet in twenty-four hours.

By using carbolic acid and other disinfectants, in the room where the first case occurs, and in and around the house, the disease has usually been stopped. By using it in streets that bound the square where the yellow fever existed, it seldom went beyond that square.

Carbolic acid disinfectant was first systematically resorted to in New Orleans in 1870. In every instance the evidence indicated favorable results. Where it was resorted to the disease was stayed, or stopped. Where not used, in cases where the owner or occupants refused to permit it—it continued and increased.

In 1871 the prompt use of disinfectants confined the disease to a few localities, principally above Canal street; and the total number of cases in the whole city was but 114, of which 55 were fatal. Elsewhere, in every Southern town where yellow fever appeared at all that year, and disinfection was not resorted to as a preventive measure, it became epidemic.

In 1872 there were eighty-three cases and thirty-nine deaths in New Orleans; sixty-one cases originating in the Fourth District. Cases occured in each of the six city districts, but disinfection prevented it from becoming epidemic anywhere.

In 1873 the bark Valparaiso, brought the yellow fever from Cuba to New Orleans. The disease was transmitted to each of the six city districts; but again, in consequence of prompt and systematic disinfection, it became epidemic in none of them, notwithstanding its very early commencement. In the First District there were 40 cases; in the Second, 37; in the third, 36; in the Fourth, 123; in the Fifth, 34; in the Sixth, 28; and the balance in the hospitals.

There can be no well founded doubt that the yellow fever epidemic of Shreveport and Memphis in 1873 originated in New Orleans. The evidence is abundant, indeed overwhelming, that New Orleans was saved from a terrible yellow fever epidemic that year by the efforts of the Board of Health, its sanitary corps, and the use of disinfectants. Did space permit, a lengthy statistical account, in proof of this statement, might be given, as contained in the Annual Report of the Board for that year.

In the Fourth District no subsequent cases occured in fifteen out of twenty blocks disinfected that year, and very few in the remaining five blocks.

In another infected locality, of twenty-five half squares disinfected, no subsequent cases occured in twenty of them, and but ten in the remaining five half squares.

In the whole Fourth District, on shore, ninety-six persons took yellow fever before disinfection, and but twelve afterwards.

Galveston, which has repeatedly suffered from yellow fever epidemics, escaped with less than fifty cases in 1873, by adopting the New Orleans process of disinfection.

In the infected Second District, in 1875, seven cases occured in August, twenty-five in September, sixteen in October and four in November. The conditions were favorable in this portion of the city for the rapid spread of the disease; but disinfection, as usual, notwithstanding the local opposition which delayed the results, finally subdued it there entirely.

Two cases occured near the Jackson Railroad depot in September, but after disinfection there were no more.

In the Fourth District, back of Dryades, seventeen cases appeared in September within two areas, comprising seventeen blocks in all, near to each other. Then however, there was no opposition to disinfection, and in all cases the houses and yards were thrown open to the disinfecting agent.

A total cessation of fever in that district resulted immediately thereafter, notwithstanding that a census of the infected blocks established the fact that nearly six hundred unacclimated white persons were residents therein. No stronger proof of the value of disinfection could be desired. The cause and effect are unmistakable.

The Sixth District was thoroughly treated with crude carbolic acid and other disinfectants, and the disease entirely disappeared on the 27th of September. As there had been no frost, and there was no lack of unacclimated material, why was the progress of the disease arrested at this point, unless by the agencies above mentioned? Eighty-four barrels of crude carbolic acid, besides other disinfectants, were spinkled in the streets, yards, and privies of that section of the city.

Of the thirteen officers detailed by the Metropolitan Board, to assist in sanitary work and disinfecting, no one took the fever, though all save one was unacclimated. Did not the constant use of disinfectants give them immunity from this disease? They were all exposed to it, day after day, for nearly five months.

It is believed that but for disinfection yellow fever epidemic would have scourged New Orleans in 1875.

Since Mobile and Galveston have adopted the system of disinfection, they too have escaped.

Yellow fever was not known in New Orleans prior to 1796.

Dr. C. B. White, President of the Board of Health, thinks the disease might become extinct in New Orleans if not reinforced from localities where it is epidemic and perennial.

From 1812 to 1833 there were twelve yellow fever epidemics in New Orleans, an average of four epidemics in seven years.

From 1833 to 1855 there were twelve yellow fever epidemics, or at the rate of six in eleven years.

Up to 1876 there have been but three epidemics in twenty-one years.

During the years 1874 and 1875 the new method of disinfecting the holds and cargoes of vessels from infected ports by forcing into them carbolic, or cresylic acid, in the bilges, etc., of the vessels, has been in use at the Mississippi quarantine station. On no vessel so disinfected at quarantine has a case of yellow fever appeared after her arrival in port.

The above facts given by Mr. Bayley are exceedingly interesting, and appear to be highly important; but, as the yellow fever epidemics had been abating, and becoming less frequent for twelve or fifteen years before the carbolic acid experiments were applied, it is difficult to say exactly to what extent epidemics have been prevented by this new application. The success of the Board of Health in hedging in and smothering the disease in numerous localities all over the city, in different years, seems truly remarkable. The Board of Health believes that epidemic yellow fever can be completely prevented in all coming time in this city, and along this coast; and the results of late experiments seem to justify the conclusion.

THE SUGAR AND RICE CROPS OF LOUISIANA.

The following statements in relation to the Sugar and Rice crops of Louisiana taken off in the fall of 1874, we copy from the book published by Mr. L. Bouchereau. This is the eighth annual report, and, like his preceding reports, it is very complete and satisfactory.

PARISHES.	1861. hhds sugar.	1874. hhds. sugar.	1874. 230 lbs to bbl. bbls rice.
Rapides	19,537	2,050	
Avoyelles	6,121	1,147	470
Point Coupee	22,565	1,883	
West Feliciana	5,712	315	
East Feliciana	716	17	
Livingston		7	
St. Tammany	60	60	
East Baton Rouge	10,949	2,358	
West Baton Rouge	24,697	2,379	

Iberville	41,921	7,303	172
Ascension	30,722	11,875	479
St. James	34,227	13,519	4,077
St. John the Baptist	18,843	7,878	9,105
St. Charles	18,191	5,289	14,400
Jefferson	11,086	3,205	2,340
Orleans	1,820	883	130
St. Bernard	6,610	2,200	4,085
Plaquemine	22,433	7,778	38,199
Assumption	37,766	11,885	1,629
Lafourche	29,781	10,251	26,460
Terrebonne	28,839	9,005	1,988
St. Mary	48,779	7,845	129
Iberia (new parish)	——	1,979	
St. Martin	16,088	1,388	
Vermilion	907	304	1,100
Lafayette	1,348	84	200
St. Landry	7,982	1,081	
Cistn' bott's on 389,264 hhds ('61–'62)	11,679		
	454,410	116,867	104,963

Estimated at........................ 528,321,500 lbs sugar in 1861
134,504,691 " " " 1874
11,516,828 gals molasses " 1874
60 gallons molasses to a hhd sugar.. 27,564,600 " " " 1861

GENERAL REMARKS.

There were 92,069 acres of sugar cane ground in Louisiana in 1874 which yielded an average of 1,462 pounds of sugar, and 125 gallons of molasses, to the acre. This large proportion of molasses was caused by the good prices paid for molasses, and by the early freezes of the 1st and 2d of November, which so injured a portion of the cane that the juice was converted into syrup without any attempt to convert it into sugar.

In making up the total amount of the crop in pounds we have adopted 1137 pounds net per hogshead for the brown or kettle sugar and 1221 pounds net for the refined, that being the exact average weights of last season.

15

The average number of gallons of molasses was much larger than usual, 101 gallons per hogshead of sugar. The old reckoning was usually about 60 gallons of molasses drained from a hogshead of sugar in ordinary seasons.

SUGAR HOUSES.

The number of sugar houses in operation in 1874 was 1000, of which 766 used steam, and 224 horse power. In 1873 there were 1080 sugar mills in operation in the State, of which 856 used steam, and 224 horse power.

Of the sugar-houses in operation in 1874, 948 produced 97,499 hhds of brown sugar, including cistern bottoms, weighing 110,856,363 lbs., and 52 sugar houses made 19,368 hhds clarified sugar—23,648,328 lbs.

CONSUMPTION OF SUGAR AND MOLASSES.

The United States consumed, in 1874—

Foreign sugars	661,869	tons.
Domestic sugars	48,500	"
Total	710,369	"

Consumption of molasses, same year—

Foreign	39,506,267	gals.
Domestic	8,699,990	"
Total	48,206,257	"

LOUISIANA SUGAR CROPS—FROM 1823 TO 1874—50 YEARS.

Crop of			*Crop of*		
1874	116,867	hhds.	1849	274,293	hhds.
1873	89,498	"	1848	220,000	"
1872	108,520	"	1847	240,000	"
1871	128,461	"	1846	140,000	"
1870	144,881	"	1845	186,000	"
1869	87,090	"	1844	200,000	"
1868	84,256	"	1843	100,000	"
1867	37,647	"	1842	140,000	"
1866	41,000	"	1841	90,000	"
1865	18,070	"	1840	87,000	"
1864	10,387	"	1839	115,000	"

1863	76,801	“	1838	70,000	“
1862	no data.		1837	65,000	“
1861	459.410	“	1836	70,000	“
1860	228,753	“	1835	30,000	“
1859	221,840	“	1834	100,000	“
1858	362,296	“	1833	75,000	“
1857	279,697	“	1832	70,000	“
1856	73,296	“	1829	48,000	“
1855	231,427	“	1828	88,000	“
1854	346,635	“	1827	71,000	“
1853	449,324	“	1826	45,000	“
1852	321,934	“	1825	30,000	“
1851	236,547	“	1824	32,000	“
1850	211,201	“	1823	30,000	“

RAILROADS IN LOUISIANA.

	Miles.
Mileage of railroads in 1862	335
“ “ “ 1872	539
N. O. Mobile and Texas Railroad—Mobile to New Orleans	140
The Morgan, Louisiana and Texas Railroad	80
N. O. St. Louis and Chicago Railroad:	
New Orleans to Osyka, the State line	88
“ “ Canton	206
“ “ Cairo	548

POPULATION.

The population of the State was, in—

1810	76,555
1820	152,923
1830	215,739
1840	352,411
1850	517,762
1860	708,002
1870	726,915

RACE, DIVISIONS, 1870.

Whites	362,065
Colored	364,210
Chinese	71
Indians	569

According to their nativity, the inhabitants in 1870 were thus divided:

Born in the United States	655,088
Natives of the State	501,864
Born in foreign countries	61,827
British Americans	714
Germans	18,933
English	792
Irish	17,068
Scotch	814

ABOVE 10 YEARS OF AGE.

According to the census, in 1870—

Population of the State, 10 years of age and over	526,392
Engaged in useful occupations	256,452

As follows:

Agriculture	141,467
Professional and personal services	65,347
Trade and transportation	23,831
Manufactures and mechanical industries	28,807

NEW ORLEANS.

New Orleans is situated in latitude 29° 58' north latitude. The streets are several feet below the surface of the river in the highest freshets, and are protected by levees.

In 1722 it had 100 wooden houses, and about 200 inhabitants.

In 1800 the population was	8,000
" 1820 " " "	27,176
" 1840 " " "	102,193
" 1860 " " "	168,823
" 1870 " " "	191,418

The population comprises—

Natives of the United States	142,943
Born in Louisiana	78,209
Natives of Germany	15,224
" " Ireland	14,693
" " England, Scotland and Wales	2,643
" " France	8,806
" " Spain	951
" " Switzerland	668
" " Cuba	775

DISTANCES.

	Miles
New Orleans to South West Pass—river	100
" " " " air-line	80
Upper line of Louisiana to S. W. Pass	590
Air-line to S. W. Pass	300

New Orleans to Donaldsonville	80
" " Plaquemine	110
" " Baton Rouge	130
" " Port Hudson	150
" " Bayou Sara	165
" " Mouth of Red River	220
" " Natchez	280
" " Vicksburg	400
" " Louisiana Line	515
" " Mississippi Line	780
" " Memphis	800
" " Arkansas Line	915
" " Tennessee Line	970
" " Cairo	1050
" " St. Louis	1250

New Orleans to Fort Jackson	75
" " Balize	100
Air-line upper line State to Balize	300
By river	615

THE MISSISSIPPI RIVER.

The length of the Mississippi river from its source to the mouth of the Missouri, as is stated in Capt. A. A. Humphrey's report, is—

	Miles.
	1330
From the mouth of the Missouri to the mouth of the Mississippi	1286
	2616

It rises in Lake Itasca, in latitude 47°, and empties into the Gulf of Mexico in latitude 29°.

	Distances from Mouth. Miles.	Height above the sea. Feet.	Fall per mile.
Extreme source	2,616	1,680	
Lake Itasca	2,610	1,575	17.50
St. Paul	1,944	670	2.92
Prairie du Chien	1,739	600	0.34
Mouth Missouri river	1,286	416	0.40
Cairo	1,097	322	0.49
Memphis	872	221	0.45
Baton Rouge	245	34	0.29
Gulf of Mexico	0	0	0.23

The Mississippi river, with its tributaries, drains an area of 1,244,000 square miles. Its average depth below the mouth of the Ohio varies from 90 to 120 feet.

N. E. LOUISIANA—OVER THE LAKE.

The following article on North-eastern Louisiana, or that portion of the State that lies north of Lake Pontchartrain, we take from the Agricultural Department of the N. O. Times. It was written by Col. Frank Bartlett, the agricultuaal editor of that journal. Col. Bartlett has been connected with the New Orleans press for more than twenty years, is an able and vigorous writer, and takes a deep interest in settling up the State with a solid yeomanry. He is intimately acquainted with the topography and general interests of the State, and particularly with the section of Louisiana which he so well describes in the following article:

BOUNDARIES—ELEVATION ABOVE TIDE WATER—LAKE PONTCHARTRAIN.

It will no doubt be interesting to our readers to learn some thing about this section of the State—that part of the Commonwealth bounded by the Amite River on the west, the Pearl River on the east, lakes Pontchartrain and Maurepas on the South, and the State of Mississippi on the North. Our limits will not admit of a close topographical description, but it will serve every purpose to give, as we design, a general view of the country and its capabilities from an agricultural, mechanical and commercial standpoint.

Lake Pontchartrain, which washes the greater part of the southern border of this territory, has an average width of twenty miles. There is a narrow rim of swamp on its north shore—a mile in width at its extreme eastern end, and widening out to six or seven miles at the point where the Amite River enters Lake Maurepas. North of this swamp line the land gradually rises until it attains an altitude of nearly two hundred feet at the Mississippi State line, immediately north of New Orleans. The elevation, however, is not so great further west, as it is only 88 feet at Osyka on the Jackson Railroad.

For the first fifteen miles as you go north from the lake, the face of the country is not marked by any sudden elevations, but appears a gently undulating plain, covered, where the timber has not been cut off for lumber or fuel, with a dense growth of long leaf pine, or on the borders of water courses with oaks of various varieties, gum, magnolia, hickory, persimmon and sweet gum.

SOIL AND SUBSOIL.

There is a thin strata of sandy loam upon the surface, underlying which is a compact clay subsoil, that varies but little in quality throughout the whole region. So plentiful is this clay, and so valuable for brick making, that the manufacture of this building material has been the chief pursuit of the population since the first settlement of the country.

This clay being tenacious, and so generally underlying the surface, the lands are capable of being fertilized to any extent, and when plowed deep the clay also contains certain elements of plant food which contribute liberally towards assisting the farmer to recover from his ground a reward for his labor.

PURE SPRING WATER, AND NUTRITIOUS GRASS.

In this gently undulating region, springs of valuable mineral qualities, as well as those of pure water, abound; and every few miles it is intersected by small rivers, creeks and branches, all fed from these sources and affording easy and reliable navigation, as well as a constant and plentiful supply of pure water for stock. The face of the plains is covered throughout the year by a nutritious grass, sufficient, were it not wantonly burned off, to supply abundant forage for all the cattle in the State the year round.

HILLS—TIMBER—CREEK BOTTOMS.

Further north still—say sixteen to twenty miles above the lake on the eastern border, and twenty-four to thirty further west, begin the hills. Commencing by gentle elevations, they rapidly grow higher and higher, with a constantly increasing elevation of the general face of the country, until the Mississippi State line is reached. The soil is deeper, the clay not quite so stiff, and the reward of the agriculturist greater. The growth of timber is also larger, both in the uplands and bottoms; and the pasturage to be surpassed only in the prairies of the Great West. The lands on the rivers and creeks are exceedingly fertile, producing remunerative crops of cotton, corn, rye, barley, oats, sugar cane and sweet potatoes. The hill lands, when properly cultivated and moderately fertilized, pay well, and by many are preferred to the bottoms, being lighter and consequently requiring less labor in their cultivation. This description will generally answer for all the country comprised in the parishes of St. Tammany, Washington, and the eastern portion of Tangipahoa. That in the parishes of St. Helena, Livingston, and the western portion of Tangipahoa contains richer lands, less pine, and a very heavy growth of oaks of various kinds, magnolia, gum, sweet gum, etc., etc, such as are usually found on hammock lands elsewhere; and while the country is better agriculturally, it is not so salubrious nor agreeable as a place of residence. Still, on account of its close proximity to railroad communication, it offers great inducements to the emigrant.

RIVERS—BAYOUS—CREEKS.

The rivers, bayous and creeks which intersect this region are numerous—too numerous for us to mention in detail. Many of them are navigable at all seasons of the year, affording a cheap means of getting products of all kinds to market. The Pearl River, which empties into the Rigolets, between Lake Pontchartrain and the Mississippi Sound, is navigable to points above the Mississippi State line; the Tchefuncta and Bogue Falia, to Covington at all seasons; the Tangipahoa, for a considerable distance in the interior; the Natalbany to Springfield, and the Amite to the State line. All these rivers have pure clear water and are filled with fish, and there are good mill sites, affording ample water power, every few miles, upon most of them.

GRAPES—FRUITS—PECANS.

Besides the products usually grown in this State, this section appears to be peculiarly adapted to the cultivation of fruits and grapes. The pecan nut is indigenous and attains greater perfection than even in Texas, while we have no doubt but that if the walnut, chesnut and shell-bark were attempted, the experiment would prove successful.

SUGAR-CANE—TOBACCO—UPLAND RICE.

Sugar cane has been grown very successfully, and when the lands were well fertilized, the product has fully equaled, if not surpassed, the best alluvions of the Mississippi Delta. The cane appears to be richer in saccharine juice, while the labor of cultivation is not nearly so great. Tobacco and upland rice also do well there, and the *morus multicaucis*, the mulberry upon which the silk-worm feeds, grows almost withont cultivation. Indeed the experiment has been made in producing silk, and it has proved very successful.

BRICKS—TILE AND POTTERY.

As we mentioned above, the supply of clay for brick, tile and pottery manufacture is inexhaustible and of the most superior quality. The numerous rivers afford ample transportation of the cheapest kind for these things, while the almost limitless forests supply fuel in any quantity for their burning.

HEALTH AND LONGEVITY.

In point of salubrity this section is not to be excelled by any part of the Continent. Coming through such a long stretch of pine forests the breezes are eliminated of all miasma and laden with resinous balm. Yellow Fever, cholera and the other plagues have never taken a foothold there; ordinary fevers and rheumatic complaints are rare, and we will venture to assert that there are more very old people, according to population, than in any locality in the South. Tha temperature, winter and summer, never verges on the disagreeable, nor reches a point when labor, out of doors, is dangerous to the health.

GRASSES AND SHEEP RAISING.

The vast ranges covered with nutritious grasses give this country a great advantage for cattle and sheep raising. In the last, particularly, are there to be realized great profits. Sheep keep fat the year round without the husbandman having to supply them with an ounce of food. Another advantage is, there are no burs, and consequently the wool is cleaner and more saleable.

GAME.

Game of all kinds is abundant, while animals of a destructive nature are very scarce and only to be found in regions remote from population—in swamps or the heavily timbered river bottoms. Beavers and otters also are found in great numbers, and the sale of their furs form no inconsiderable addition to the earnings of the people each year.

POPULATION.

The present population of this section of the State consists chiefly of natives of Georgia and South Carolina and their descendants; on the Jackson Railroad and in the country adjacent to it are to be found many families, who formerly resided in New Orleans, but who have embraced agriculture *con amore.* In the towns and villages will be found about the same class of society as in other parts of the State; but judging from the fact that the bar and billiard-rooms do not appear to thrive, we take it that the love of gaming and drinking—the besetting vices of the Anglo-Saxon—do not rule to a great extent among those people.

PRICE OF LANDS, ETC.

In conclusion we will state that, in nearly all of the parishes of the section we have been describing, there are vast bodies of unoccupied lands awaiting settlement. A good part of these lands belong to the United States, and can be entered as homesteads or bought at the usual government price. The land owners are willing to sell at low rates to actual settlers, and will welcome any honest and industrious farmer who comes among them, with open arms and true old fashioned Georgia hospitality. But whoever goes there must expect to work hard, as it is no country for genteel farming or kid glove gentility.

A LIVELY DESCRIPTION.

The following article from the Co-operative News of the 18th of

November, 1875, written by its editor, Judge John B. Robinson, when the remembrance of all the minute particulars were fresh in his memory, includes so many points throwing light upon some of the peculiar customs and scenery of Southern Louisiana, and all so well said, that we cannot resist the inclination to publish the entire article. It is certainly an interesting narrative, in a racy style, that gives relief from the monotony of facts and figures. It is one feature of life among the lakes, and bayous, and marais of Louisiana, well delineated by the artist's pen.

LES CHASSEURS DE CANARDS—A DAY AMONG THE DUCK HUNTERS.

A few days since business called us to Lafourche, and an opportunity offering, we were tempted to spend a day and night with the duck hunters of Lake Le Bœuf.

Under the guidance of our friend Cyprien Mathern we were equipped with a gun and ammunition, two blankets, and a sack with ample supplies of coffee, sugar, salt, pepper, biscuit and potatoes. All of which we shouldered, and marching off to the rear of the rice fields for a mile and a half. there took a light pirogue, and followed Cyprien in his. We paddled through canals and bayous, through dark cypress swamps and reedy prairies, and in about two hours reached, by 1 o'clock in the day, the destined camp, a rough but roomy hut, built of broad cypress pieux, or pickets, sides, roof and floor of the same material. A large shed, close at hand, served as the kitchen or cook house. This camp was located about five hundred yards from the open water of the lake, upon a canal. It stood in the midst of a floating prairie, or *prairie tremblante.* Some tall willow trees had grown up here, and formed an island of firm land; all around was floating prairie, a foot to two feet thick, formed by the matted roots of the rank grass, and the accumulations of ashes and decayed vegetable matter, so light as to float, yet firm enough to support a man, while underneath the water was clear and pure, and three to four feet deep.

This floating prairie covers many thousands of acres of what was twenty years ago clear water five or six feet deep; it is constantly encroaching on the lake, and thickening, and will eventually settle down to the bottom and form a prairie just as the other marsh-prairies of Louisiana were formed. The vegetation equals in rankness anything we have ever seen, and each season a loose mat of tangled reeds and grass two feet thick is formed. One of the principal grasses is the wild oats, *folle avoine*, ten to fifteen feet high, whose rich, farinaceous seeds furnish food for rails and wild fowl, and would easily furnish food for man, as the Indians found it wholesome and abundant.

This prairie fairly swarms with rails of every variety, *poule d'eaux* or water hens, and swamp rabbits almost as large as a kid. When the frosts have killed the grasses this prairie is burned off, and then the hunters come with their dogs and for many weeks can kill from twenty-five to a hundred in a day.

Lake Le Bœuf is about six miles long and three broad, and has been a famous ducking ground ever since the country was settled.

There are ten or twelve camps, each containing five to six hunters; these hunters form a partnership, running from October to March, and maintain a steady camp during the whole fall and winter, for the purpose of hunting for the market. They are all Creoles or Acadians, originally from the French settlement of Acadia, in Canada, about whom Longfellow wrote his poem of "Evangeline." They send out their game every day to the Morgan Texas Railroad and ship it to their salesman in New Orleans, who keeps an account current for the hunting season with the recognized head of the partnership.

Our camp consisted of four hunters, with several independent sportsmen. We found them preparing their dinner, which invariably consists of water hens, or *poule d'eaux*, and boiled rice, highly seasoned and called a *jombalyeeah*, and a pot of boiled coffee; they use no bread whatever, but occasionally roast a few sweet potatoes.

After dinner the guns, all double-barreled muzzle loaders, were carefully cleaned and examined; the shot, No. 7 only, and powder replenished; and dry, black moss for powder wadding, and green moss for the shot, were placed in convenient reach in the pirogues, and soon each one, with gun in front of him and paddle in hand, was off for his evening hunt.

This evening hunt consisted in cruising cautiously and silently around the borders of the open water, peeping into the little nooks and around the little islands of floating prairie, chiefly for *poule d'eaux*; but also for ducks, and the banging of the guns told that execution was going on rapidly on all sides. All these hunters are capital shots and take their game on the wing, dropping their paddles, snatching up their guns and dropping the game when flushed, with astonishing expertness. Toward sunset all ensconced themselves in blinds made out in the open water, by sticking into the shallow water long green reeds, in such a manner as to leave the tops four feet above the water, in two rows, the length of a canoe, with one end closed and a space left between just wide enough to admit a pirogue, or canoe. Into one of these blinds the hunter, with a duck call, stations himself and awaits the flight of the ducks as they pour into the lake toward nightfall. From an hour before sunset until after dark the flashes and booms from these treacherous blinds tell the destruction that is going on.

When it was too dark to shoot with accuracy, the hunters turned their pirogues toward camp, and upon their arrival we found that they had from fifteen to thirty each of ducks and water hens, but chiefly ducks.

Upon their arrival at camp one of the professional hunters makes the fire and puts on three pots of water—one to scald what fowl they design to eat, one for coffee, and one for the invariable rice and water hens.

The others lay the ducks in a pile; one picks up a duck, turns its belly up, strains back the wings, plucks at a single stroke a large handful of feathers from off the side below the wing, and pushes the feathers into a bag hung before him; snatches another bunch from off the other side, then two from the lower part of the belly, leaving a bare place about the size of one's palm. He then tosses the duck to another man, who is sitting on the floor; he carefully and cautiously opens the belly with a small narrow bladed knife, very sharp, and then throws the duck to a third man who draws out the entrails with a quick jerk with one hand, and pulls off the liver, gizzard and heart at a grasp, throwing them into a bucket, to be dressed and eaten in camp. Another takes up the duck, presses into the cavity a ball of green moss, which he rubs around, withdraws and throws away, putting in as many as three or four of these moss balls until the bird is clean and dry inside. The ducks are then sorted, tied in pairs and hung up by the necks until packed to be sent off.

No less than sixteen varieties of ducks are known to these hunters. The canvas back and its congener, the red head, are the best, the black duck is next, and the French duck or mallard is third, or even fourth with some, as the large cheval or horse duck is considered as good or better than the mallard

The camp at night is an interesting picture. When the ducks are disposed of each one takes a poul d'eau and plucks off the feathers as clean as possible; the bird is then dipped into scalding water and rolled in hot ashes, when it may be rubbed down white and clean. The entrails are then taken out, it is washed and a small stick about two feet long is stuck into its mouth, passing along the skin of its neck and out through the rear of the body. Onions or garlic, and salt and pepper are then pressed into it and the stick is stuck into the ground in front of the fire at such an angle as to make the bird hang over the fire. In the course of an hour, by occasional turning, it is roasted to a turn, when the hunter sticks the stick in the ground between his legs, and pulls off wings and legs, and, finally, the breast and body at his leisure; and it is certainly as rich a feast as one would wish to eat. Then comes the jombalyeeyah and the coffee, when pipes, songs and stories in Acadian French make the fireside merry for a couple of hours.

By nine all is quiet, and nothing is heard but the incessant hooting of the owls in the distant cypress swamps, with the call of the rails and water fowls when disturbed by the rabbits.

Four o'clock brings up the camp to a cup of strong coffee, and then before day all are on their way to their blinds, and by early dawn the cannonading from sixty hunters sounds like a battle.

Just as the sun rises millions of swallows suddenly make their appearance, and for an hour skim low down over the lake, gradually rising higher until they suddenly take flight and all are gone.

MOSS.

THE MOSS OF COMMERCE—HOW IT GROWS—WHERE IT GROWS—HOW GATHERED AND CURED—EXTENT OF THE INDUSTRY—PRICES AND MARKETS.

The following article from the pen of Judge John B. Robinson, editor of the Co-operative News, and author of the "Resources of Louisiana," will be read with interest. Spanish moss is an article of no small importance to Louisiana, and the facts given below contain many features of rare interest:

The long moss of our commerce is almost wholly a product of Louisiana; although found in the low lands of South Carolina and along the swampy borders of all the Gulf States, it is no where else found in sufficient quantities to make it an object of commerce.

It is rarely found above 33 degrees north latitude, and usually keeps company with the palmetto.

Its technical name is *Tillanitsia Usnevides.* It is commonly called Spanish moss, long moss, and gray moss. Its native habitat is on the tops and branches of living trees which grow in the gloomy swamps or along their borders. It revels in the darkest recesses of the deep and dismal cypress groves, above the exhalations of everlasting swamps, and covers as with a mantle the broad-armed live and native oaks which fringe the ridgy margins of the lakes and bayous.

It even drifts away from the tops of the cypress and tupelo, and encroaches on the highlands adjacent to the swamps, and festoons with its gray drapery, the sweet gum, elm and ash.

Associated as it is by false report and preconceived ideas, with malarial fevers and swamp ague, the stranger, when he first views the long pendulous pennants of the gray moss, solemnly swaying in the breeze, cannot resist the impression that he is looking on the waving plumes of a hundred hearses.

But prone as the imagination is to this delusion, it is now well settled that this long moss is the salvation of the swamp residents.

Many a home along the dark margins of extensive swamps enjoy as perfect health and as great immunity from disease as those do which are located in the mountains.

This moss needs the tree simply too keep it in the air. It is, therefore, an epiphyte. It is not a parasite, because it does not derive any sustenance from the tree; but it feeds on the malarious elements in the atmosphere, and, consuming them, purifies the surrounding air, which would, for human lungs and skin, be otherwise loaded with poison, from the rapid decay of exuberant vegetation.

It cannot live on a dead tree, because the bark, among the crevices of which its tendrils creep, has slipped off. When the tree dies, the moss soon turns black, and drapes itself in mourning, as if for the tree, its dead mother.

No scenery in nature can convey a more solemn and impressive feeling to the traveler than a moss covered swamp. As he pushes his pirogue through the lofty wreathes and verdant arches of the silent swamp, the tall columns of cypress rise up on every side like huge stalagmites, upholding the leafy, living cavern above, from the roof of which depend long masses of moss, like innumerable gray stalactites, so shutting out the sun as to make it twilight at noon.

The living moss is of a greenish gray color. It has long branching fibres or filaments, and at each bifurcation produces tiny, trumpet-shaped flowers, smaller than tobacco flowers, and of a peach blossom color It grows rapidly, and is easily propagated; a single thread blown from one tree to another, soon grows into a mass of moss.

The great moss region is all Southwestern Louisiana, wherever swamps exist, from Alexandria, on Red River, to the Sabine, and from Alexandria, down Red River to the City of New Orleans. This includes the Atchafalaya basin and Teche; all the main lakes and bayous with the Grosse Tete, Plaquemine, Lafourche, Terrebonne and Barataria regions.

THE REGION OVER THE LAKE

also furnishes a considerable yield of moss; this region includes all East Louisiana, Lower Mississippi, Alabama and Florida.

THE BEST MOSS COMES

from the Atachafalya basin. It is long, soft, fine threaded and glossy; this moss is the product of cypress swamps alone.

That from Lafourche, Terrebonne and Bayou Black is coarse, gummy, hard to clean, but very strong. It is gathered largely from gum trees and seems to be glued with the gum.

It is gathered by wood choppers and laborers who follow the wood choppers, but there are hundreds of whites and blacks who make it a business. They go into the swamps through canals and bayous; they push their way along in skiffs, flats or canoes; they carefully pick up all that the wind has blown down in great flakes; they reach up long poles, armed with hooks, and pull down the hanging bunches, and they, if necessary, climb the trees and throw the moss down. It is piled up in heaps, if on highland, where it is gathered, or if in the swamps, it is brought out in boats and piled in convenient heaps of several hundred pounds, like hay ricks or shocks.

When cured to the owners satisfaction, it is scattered and dried, it is then hauled or sold to the country dealers who bale it. Country moss is baled like hay in bales of unequal size, weighing from 200 to 500 pounds weight.

It is baled by horse or hand power, by screw presses and by lever presses, the latter being usually a long log put into a mortice in a tree. These bales are all usually very rough and often contain old collars, matting and such articles. They are fastened by cypress split boards on four sides, and bound with wire, or wooden hoops or bands, and often with vines.

A good hand can gather enough in a day of green moss to put up 250 to 500 pound bale of black moss. When baled it is sent to the city dealers, who either buys directly or sells on commission to purchasers or speculators.

There are some half dozen different firms in this city engaged in receiving on consignment and selling moss.

These receive consignments from country dealers and original shippers. They also purchase all unbaled lots which may come in from across the lake, or the plantations near the city.

Like cotton going to the cotton press, no matter who is the purchaser from the city dealers the moss all goes from the hands of the city dealer to the

CITY FACTORIES.

There are but two factories, and all moss to be saleable must have the brand of one of these factories.

By cleaning and preparing the moss and making uniform classification, the moss trade has largely increased under the factory system.

The business of the factory is to open up the rough bales, sort the moss and gin it, so as to get rid of leaves, sticks, dust and trash; it is then boiled by steam in a secret bath; but the secret may easily be guessed. It is supposed to be simply copperas or sulphate of iron, in the water, which, when boiled with the moss fixes the tannin in the moss and turns it into a deep glossy black.

After boiling, the moss is dried, sometimes reginned, and then classified into four grades as follows: X gray, XX gray and brown, XXX brown and black, XXXX all black.

These grades are all uniform to both factories. It is put up in burlap bales of 100 to 120 pounds, called quarter bales, and 175 to 200 pounds, called half bales. This is known as machine picked moss. Very little hand picked mosses sent out of New Orleans—not over 1500 bales annually.

These factories each employ from 10 to 15 men, and from 15 to 20 women, and each turns out from 30 to 40 bales a day. They are about equal in their operations, and together have turned out during the past year 19,000 bales.

Some moss is shipped from Berwick's Bay to Texas, but not much—not exceeding 800 bales per year.

The receipts at New Orleans average about 600 irregular bales per week. The busiest season is from middle of January to June, and slacks off from November to January, owing to the freezing up of Northern factories.

The prices are for green moss in the country, from 25 to 50 cents per hundred, for cured moss in the country 1 to 1½ cents a pound. City dealers pay for:

* Gray,	2@3	cents ℔ ℔.
** Gray and brown	3@4½	..
*** Brown and black	3@5	..
****All black	4@5½	..

Crescent or Delta brand of machine picked moss sells at

* Gray	7	cents ℔ ℔.
** Gray and brown	8	..
*** Brown and black	9	..
**** All black	11	..

The total sales of the past year have been 20,800 bales. This can only be estimated as the bales are irregular. It must be borne that green moss when piled loses a large per cent. of its weight, and that one pound of ginned clean moss takes 7 of green moss.

Moss is sent to all parts of the United States and Canada, and large quantities are sent to France and Germany. It is used for making mattresses, stuffing chairs, cushions, car seats and all uses to which hair was applied.

PERIQUE TOBACCO.

The best of Perique tobacco is made only in St. James parish. [It has been made in Natchitoches, Chachahoula, and lately in St. Tammany parish. This tobacco, when well cured and prepared for market, is very black, has a delightful odor, and is the delight of those who are fond of pipe pleasures. It is also made into snuff which old snuff-takers think delicious. It is put up in rolls weighing from three to four pounds each, wound with a cord drawn tightly until, when dry, on the outside, it is exceedingly hard and solid.

The land on which this tobacco grows is usually light, sandy, magnolia soil. In St. James the seeds are usually sown in the open air in January, covered with palmetto leaves in cold weather, and the plants set out in March, three by four feet apart.

There are about twenty families engaged in the cultivation of Perique tobacco in St. James parish. The yearly crop is usually about 20,000 carrotes of four pounds each. An acre of this tobacco usually makes about 80 to 100 carrotes, worth from $3 to $3 25.

PUBLIC LANDS.

The public lands in Louisiana belonging to the United States amount to about.................................7,000.000 acres.
Belonging to the State...............................9,000,000 "
This includes the sea marsh and lands subject to overflow.

Public lands subject to overflow belong to the State.

Homesteads, from United States lands in Louisiana, can be obtained at an expense of about fifteen dollars for a quarter section of 160 acres.

The principal part of the United States lands in Louisiana lie in the parishes of Washington, St. Tammany, St. Helena, St. Landry, Calcasieu, Cameron and N. W. Louisiana.

THE JETTIES.

AT THE MOUTH OF THE MISSISSIPPI RIVER.

The Jetties at the mouth of the Mississippi, at the South Pass, bid fair to be a success. Judging from the deepening of the channel up to the middle of February, they expect to get over twenty-five feet of water in less than six months.

DESCRIPTION.

Imagine two rows of piling, 12 feet apart, to be driven in the sand and mud at the mouth of the Mississippi, extending 2¼ miles from deep water in the river to deep water in the Gulf.

Then plank these piling on the inside, like two lines of fences of plank and posts, and fill to the tops of the posts with small willow trees and rocks.

One thousand feet east of this, the other side of the shallow channel, place two other rows of piling, filled with willows and stone in the same manner as the first. Line the inside tier on either side of the channel, with willow mattresses, two feet thick and 40 feet wide, woven and fastened together, sink them to the bottom and cover with stones, and narrower matresses, on top, to the surface of high water, by the side of the piling.

Thus, two artificial banks are formed, and between these banks, (the inside edges of the matresses), is a surface of mud and sand about 900 feet in width, and over two miles in length, for the current to play upon and wash out.

They expect, during the high water season of 1876, that this channel will wash out to a depth that a vessel drawing twenty-five feet of water will be able to come up to New Orleans.

INTERESTING AND INSTRUCTIVE TABLES COPIED FROM THE ANNUAL REPORT OF THE BOARD OF HEALTH TO THE GENERAL ASSEMBLY OF LOUISIANA, FOR THE YEAR 1873.

TEMPERATURE OF CLIMATE.

Table showing Mean and Extreme Temperature in New Orleans each month in 1873.

TEMPERATURE.

	Maximum.			Minimum.		
Month.	Highest.	Lowest.	Mean.	Highest.	Lowest.	Mean.
January	71.0	58.0	56.26	61.0	25.0	43.50
February	81.0	58.0	67.91	68.0	43.0	52.32
March	81.5	54.5	69.46	66.0	38.0	53.29
April	86.0	62.0	75.98	73.5	45.5	59.92
May	88.5	72.0	82.33	74.0	57.0	68.05
June	92.0	82.5	88.35	78.0	72.0	74.97
July	98.0	80.0	90.48	82.0	70.5	76.66
August	92.5	83.0	88.83	78.0	71.0	75.74
September	91.0	80.0	85.48	78.0	65.0	73.65
October	88.0	54.0	75.71	73.5	38.0	60.71
November	77.0	50.0	68.55	66.0	37.5	54.37
December	78.0	45.0	63.00	69.0	32.0	50.84
Year			76.028			62.001

CLEAR, FAIR, CLOUDY AND RAINY DAYS, ETC.

Table showing the Number of Clear Days, Fair Days, Cloudy Days, Rainy Days, Number of Days when Lightning was observed, and Occurrences of Frost in New Orleans, in 1873.

Month.	No. of clear days.	No. of fair days.	No. of cloudy days.	No. of rainy days.	No. days when lightning was observed.	No. of occurrences of frost.
January........	6	13	5	7	1	1
February.......	5	16	1	6	1	1
March	11	6	4	10	5	2
April	10	11	6	3	2	0
May	7	8	0	16	13	0
June...........	0	12	2	16	14	0
July	5	8	1	17	14	0
August.........	4	6	7	14	12	0
September	7	4	9	10	3	0
October........	19	6	2	4	0	1
November	12	8	2	8	1	1
December......	6	7	12	6	0	4
Year.......	92	105	51	117	66	10

RAINFALL.

Table showing the Amount of Rainfall in New Orleans in 1873.

Months.	Inches.
January	5.04
February	2.20
March	5.54
April	1.38
May	21.87
June	7.36
July	7.43
August	10.46
September	3.27
October	1.43
November	7.77
December	1.38
Total	75.513

The year 1873 was remarkable for its excessive rains in Louisiana. The ordinary rainfall in this State is but from 45 to 50 inches yearly.

MORTALITY.

Tables of the Mortality in New Orleans, in 1873, *giving Color, Sex, Age, and Nativity.*

Aggregate population of New Orleans in 1873		191,418
Born in the United States		142,493
Of these, white	93,069	
" colored	48,858	
Born in Louisiana		114,686
Of these, white	78,209	
" colored	36,477	
Born in Germany		15,224
" " Ireland		14,693
" " France		8,806
" " England, Wales and Scotland		2,658

MORTALITY—1873.

COLOR.

	Jan.	Feb.	March.	April.	May.	June.	July.	Aug.	Sept.	Oct.	Nov.	Dec.	Total.
White........	388	312	328	341	568	332	523	341	457	510	342	302	4739
Black........	181	177	203	188	242	144	149	97	100	124	102	122	1829
Mulattoes....	68	62	72	54	112	76	95	54	60	72	52	78	855
Not stated...	5	5	5	17	8	10	6	6	7	7	4	2	82
Total..	642	556	603	600	930	562	773	498	624	713	500	504	7505

SEX.

	Jan.	Feb.	March.	April.	May.	June.	July.	Aug.	Sept.	Oct.	Nov.	Dec.	Total.
Males......	383	335	378	367	523	310	450	266	388	421	295	280	4396
Females....	253	217	223	231	404	248	319	230	232	285	205	223	3070
Not stated..	6	4	2	2	3	4	4	2	4	7		1	39
Total..	642	556	603	600	930	652	773	498	624	713	500	504	7505

AGES.

	Jan.	Feb.	March.	April.	May.	June.	July.	Aug.	Sept.	Oct.	Nov.	Dec.	Total.
Under 1 year.	105	85	100	117	232	134	212	120	108	145	92	89	1539
From 1 to 2 years......	26	36	28	39	92	62	68	30	23	41	33	29	507
From 2 to 5 years......	26	28	42	44	62	43	49	45	63	51	32	31	516
From 5 to 10 years......	20	33	21	35	34	15	16	14	42	29	16	24	299
From 10 to 15 years......	15	10	7	12	16	12	17	7	26	13	9	9	153
From 15 to 20 years......	21	15	14	10	26	13	19	12	19	23	8	8	188
From 20 to 25 years......	47	70	47	39	68	28	33	29	48	47	35	28	519
From 25 to 30 years......	50	42	52	39	54	30	47	23	48	60	35	31	514
From 30 to 40 years......	96	62	83	70	98	53	81	53	87	84	55	58	880
From 40 to 50 years......	67	60	81	65	75	48	80	56	57	82	50	54	775
From 50 to 60 years......	68	42	48	40	71	38	63	37	40	46	46	64	603
From 60 to 70 years......	35	33	26	33	40	35	37	31	29	36	34	35	404
From 70 to 80 years......	29	15	22	13	16	12	26	13	12	15	20	22	215
From 80 to 90 years......	9	8	7	5	7	4	7	3	5	6	9	7	77
Fr'm 90 to 100 years......	4	6	2	5	1	4	1	4	1	2	5	3	38
100 years and upwards...	0	1	2	0	3	1	2	0	0	4	1	2	16
Unknown....	24	10	21	34	32	30	15	21	16	29	20	10	262
Total..	642	556	603	600	930	562	773	498	624	713	500	500	7505

NATIVITIES.

COUNTRIES.	Jan.	Feb.	March.	April.	May.	June.	July.	Aug.	Sept.	Oct.	Nov.	Dec.	Total.
Africa........	..	1	2	1	2	..	1	2	..	2	1	1	13
Austria.......	2	2	1	1	2	1	..	3	2	6	2	2	24
Brit. America.	1	..	..	..	1	1	..	1	3	..	2	..	9
Canada.......	2	1	..	3	..	1	1	1	3	2	2	3	19
China	1	1	1	..	1	1	1	3	2	2	1	..	14
Denmark	..	..	1	..	..	1	2	3	1	..	..	..	8
England	8	6	6	10	5	4	7	6	11	10	4	5	82
East Indies...	..	..	..	..	..	..	..	..	1	..	..	..	1
France	33	20	28	14	34	16	27	12	29	33	21	23	290
Germany	49	33	32	33	52	27	53	42	62	74	40	37	534
Greece	..	1	..	..	..	..	..	1	..	..	..	..	2
Holland	..	..	..	..	1	..	..	..	..	1	..	..	2
Honduras....	..	..	..	..	1	..	..	..	..	..	..	..	1
Ireland......	62	43	52	39	51	34	57	46	57	50	50	56	597
Italy	7	5	4	4	7	1	6	3	1	2	6	6	52
Madeira													1
Mexico......	3	2	2	1	..	..	1	1	..	2	..	..	12
Norway	1	..	..	1	1	1	2	..	..	1	..	..	7
Poland										1			1
Portugal.....	..	1	..	..	..	..	2	1	..	1	1	..	6
Prussia	2	2	1	2	1	2	2	1	..	1	1	..	15
Russia.......	..	1	..	..	..	..	..	..	1	1	..	..	3
Scotland.....	6	1	2	..	3	2	3	..	3	4	1	3	28
Sea	..	1	..	..	..	..	..	..	..	1	2	..	4
So. America..					1								1
Spain	6	5	1	7	7	1	2	2	5	5	5	3	49
Sweden......		1	1	1	1			1	1	2			8
Switzerland..	2	3	2	5	3	1	1	2	5	4	1	2	21
United States.	440	442	459	470	733	457	595	555	429	480	333	350	5517
Wales										2			2
West Indies..	6	5	6	3	6	4	6	4	1	9	8	5	63
Not stated...	5	9	2	5	17	7	6	8	7	16	19	8	109
Total....	642	556	603	600	930	562	773	498	624	713	500	504	7505

MORTALITY OF CHILDREN

In New Orleans, in the Year 1873, *under* 2 *Years of Age, according to Population—United States Census,* 1870.

Month.	Number of Deaths.		Percentage of Deaths to population.		No. of Deaths to 1000 pop'ln.	
	White.	Colored.	White.	Colored.	White	Col'd.
January....	80	51	.01145	.02185	11.450	21.850
February...	70	21	.01002	.02185	10.024	21.850
March.....	65	63	.00931	.02700	9.307	27.000
April......	104	52	.01489	.02223	14.890	22.230
May.......	211	113	.03021	.04849	30.210	48.490
June......	107	89	.01532	.03813	15.320	38.130
July.......	176	104	.02520	.04456	25.200	44.550
August....	100	50	.01432	.02144	14.320	21.440
September.	85	46	.01217	.01971	12.170	19.710
October....	122	64	.01747	.02742	17.470	27.420
November..	77	48	.01120	.02056	11.200	20.565
December.	66	52	.009450	.02228	.9450	22.280
Average..	105.25	65.25	.01507	.02796	15.070	27.96
Total....	1236	783	.18156	33547	181.56	335.47

White population, under 2 years.................. 6984.

Colored population, under 2 years................ 2334.

A LIST

—OF THE—

FARMS, PLANTATIONS, WILD LANDS, TIMBER LANDS, ETC.,

—IN—

THE GULF STATES

For Sale on Commission, by the

Southern Land Company,

W. P. FRERET,

REAL ESTATE BROKER,

No. 8 Commercial Place, New Orleans.

An Arpent.—An Acre.—A surface of land 70 yards square is 60 yards more than an *acre;* 64 yards square is 8 square yards 4 morethan an *arpent.* An acre is about one arpent and one-sixth of an arpent.

An arpent is 4,088 square yards.

An acre is 4,840 square yards.

Parishes in Louisiana are the same as counties in the other States.

Plantation cabins are often good frame buildings, whitewashed, with front gallery and two or more rooms; better than most of the log cabins of the early frontiermen.

No. 20. 8 small tracts—122, 123, 108, 115, 170, 188, 213, 105 acres, Parish of E. Baton Rouge; fine, high, rich soil, and good neighborhood; fine place for a small colony.

No. 122. 213 acres, a tract of high and low land on Bayou Sara Road, Parish of E. Baton Rouge, Scotland Plantation.

No. 15. 238 acres, on Redwood Creek, E. Feliciana, 5 miles from Jackson, La., and 7 miles from Clinton : good upland, in fine locality ; good neighborhood, good water, convenient to market.

No. 2. 240 arpents, a small dwelling and 3 or 4 outhouses ; fine land ; may make a good rice farm ; located in the Parish of Ascension 5 miles above Donaldsonville, on W. bank of Mississippi.

No. 33. 300 acres, Parish of Iberville, 5 miles above the town of Plaquemine: W. bank of Mississippi river; a good brick sugar house in good order, with new cane shed: land all under fence; an old dwelling, 6 cabins and a stable.

No. 10. 308 acres fine cotton land, fronting on Little river 2 miles from Trinity; cottage residence, 3 double cabins: other improvements; neat and productive farms.

No. 58 320 arpents, Parish St. Bernard, 12 miles below New Orleans, E. bank of Mississippi river: very neat rice farm; fine soil for rice, oranges, and sugar culture.

No. 62 A——320 arpents; a very neatly improved farm on the E. bank of the Mississippi river, in St. Charles Parish; land good for cane, rice, sugar, oranges, etc., etc.,

No. 63 328 acres, Parish St. Tammany, near W. Pearl river, located in a very healthy region of country; fine water; first-class fruit landsplenty of timber for building and fencing.

No. 31 350 arpents, 12 miles below Plaquemine, nearly opposite Bayou Goula, fronting 15 arpents on the Mississippi; one small cottage dwelling and outhouses; fine shady yard; rich, sandy sugar lands, easy of cultivation.

No. 36. 370 acres, Iberville Parish, on Bayou Gross Tete, west bank; fresh rich sugar lands, with a lot of cabins and fencing for 200 acres.

No. 93 400 acres: a sugar plantation on both sides of Bayou Terrebonne, Parish of Terrebonne.

No. 70C. 400 arpents, Parish W. Baton Rouge, on west bank of Mississippi river nearly opposite Port Hudson.

No. 34. 479 acres, Parish of Iberville, very rich sugar land and fresh; one neat and comfortable dwelling, 6 rooms; outhouses, stable, 20 double cabins; beautiful shady and ornamented yard; one old sugar house burned down, machinery and bricks good; 20 miles from Baton Rouge by railroad, 25 miles above Plaquemine. Can be made a fine stock farm.

No. 13. 512 acres, good cotton and corn land; Parish of East Feliciana, good location for practical farmer; good for raising sheep and farm stock; good neighborhood; fine water; healthy.

No. 53. 520 acres, on Bayou Bœuf, Parish of St. Landry; high, rich land in good neighborhood; good for cotton, sugar, etc.,

No. 51 B. 522 arpents, a cotton plantation in Rapides Parish, on BayouRapides; dwelling, outbuildings, cabins, etc.,

No. 49. 528 acres. Parish of Rapides, 20 miles back from Alexandria; good cotton plantation, and in a good section of the Parish; good dwelling, laborers houses etc.,

No. 32. 540 arpents, Parish of Iberville, fronting on the Mississippi river, six miles below the town of Plaquemine; fine land: makes first-class kettle sugar.

No. 29. 600 arpents : sugar estate in Iberville Parish : land good for both sugar and cotton : fronting on Bayou Plaquemine. 5 miles below Plaquemine Village. a mile or two from the N. O. and Texas Railroad.

No. 8. 600 acres first-class cotton lands 1 mile below Trinity, fronting 20 arpents on west bank of Black river ; neat residence ; fine orchard and shady yard : a stable and 10 new cabins; large steam gin, two double eagle stands; 400 acres fenced.

No. 9. 600 acres of land, 1½ miles below Trinity, fronting 20 arpents on west bank of Black river, 1 neat residence and outhouses, 10 new cabins: 400 acres under fence and cultivation.

No. 82. 612 arpents, Parish of Jefferson, a plantation on Faist Island of Barataria, on Bayou Barataria.

No. 70 A. 639 acres, unimproved, in Parish West Baton Rouge

No. 67. 640 acres Tensas Parish, 15 miles west of St. Joseph; lands unimproved; fine cotton lands; can be easily reclaimed and put in cultivation; about 200 acres have once been cleared and put in cotton.

No. 46. 640 arpents; Parish of Point Coupee, 3½ miles above Bayou Sara, fronting on the Mississippi river; dwelling house and 5 rooms cabins, etc., plantation well located; can be made profitable as a dairy and stock farm: only 3 miles from the court-house.

No. 51 D. 640 acres, known as the "Yarbora Tract," section 27, T. 10R. 13.

No. 18 640 acres, Parish E. Baton Rouge, fronting on Amite river cotton plantation; good soil; 12 miles back of city of Baton Rouge; comfortable cottage house; some old cabins and outbuildings; fine timber good water and good neighborhood.

No. 16. 640 acres, 6 miles south of Clinton, half of the tract improved and half timber: a desirable an productive place; good for cotton and other crops.

No. 11. 640 acres, East Feliciana, on bank of Sandy Creek; good cotton land, can be divided into small tracts to advantage; location healthy; fine spring water.

No. 12. 640 acres, East Feliciana; good cotton or corn soil; good for fruits, and vines, and stock; healthy location; good neighborhood fine water; high and dry soil; school and church.

No. 14. 676 acres, Parish of East Feliciania; good dwelling house, and 5 good laborer's houses, stable corn crib, etc., good for cotton and corn and fine stock range; good location for profitable farming: comfortablefor a family.

No. 40. 690 acres: a productive cotton plantation in the Parish of Madison, 2 miles below Miliken's Bend, on the Mississippi river, about 30 miles above Vicksburg; 600 acres under fence: plenty of labor on the land to work it.

No. 68. 700 arpents on Mississippi river, Parish of West Baton Rouge; fine rich land well supplied with timber; good for cotton, sugar, fruits, stock, etc.,

No. 72. 713 acres, West Feliciana Parish, 8 miles back of Bayou Sara, on Woodville road; elegant dwelling, 2 stories, 11 rooms; fine schoolhouse; beautifully ornamented yard; 300 acres under fence; fine cotton plantation: fine location for a family; good neighborbood.

No. 61 A. 718 acres, Parish of St. James, west bank of the Mississippi river; a fine sugar plantation; fine dwelling, and large and lately repaired sugar house; cabin room for 50 hands

No. 59 B. 720 arpents, Parish St. Bernard, a well improved sugar plantation: new dwelling and sugar house, 10 double cabins, etc., etc.

No. 80. 756 acres, unimproved land in Carroll Parish, north of Red River.

No. 41 A. 771 acres. Parish of Madison, first class and well improved cotton plantation; 500 acres open fenced and in cultivation; one mile below Miliken's Bend on the Mississippi river.

No. 74. 782 acres, W. Feliciana, 5 miles from Bayou Sara, on the Woodville road, hill and creek bottom land; well watered and timbered; would make a fine sheep farm in connection with cotton and fruit culture.

No. 73. 800 arpents of beach flat lands, West Feliciana Parish, near Tunica Hills, 3 miles from Mississippi river: neat dwelling, 10 rooms, outhouses and cabins, a good cotton plantation; a productive hill place.

No. 53. 800 arpents, sugar and cotton plantation, Parish ot St. Landry, on Bayou Bœuf: dwelling, etc., brick sugar house in fine state of repair, with good machinery.

No. 59 A. 20 arpents front on each side Terre-aux-Bœuf's Ridge Parish St. Bernard; dwellings, outhouses, cabins, etc.

No. 27. 800 arpents, a fine sugar plantation, situated in Iberia Parish, in Fausse Point, on Bayou Teche; a dwelling, 20 new cabins, good sugar house, new fence; 400 acres under new fence, rest in wood capacity of sugar kettles 15 hogsheads of suger in 24 hours.

No. 50. 809 arpents, 5 miles west of Alexandria, Parish of Rapides comfortable dwelling; outhouses; good garden and orchard, enclosed by by a hedge; Bayou Rapides encloses the entire tract which makes a perfect drainage; first class and well improved cotton plantation.

No. 51. C. 819 arpents of fine land in Rapides Parish, on Bayou Rapides, a well improved cotton plantation.

No. 88. 830 acres in several tracts on Bullard's Plain, Parish of East Feliciana.

No. 77. 844 acres on Silver Creek, Washington Parish: good cotton land and fine stock range; part of the place improved and in cultivation.

No. 67A. 851 arpents, first class cotton and stock farm, Parish Vermillion river: dwelling, gin, and outhouses; nearly all under new fence.

No. 94. 900 acres, a sugar plantation on both sides of Bayou Terrebonne, Parish of Terrebonne.

No. 70 B. 920 and 500 arpents, 2 tracts in Parish W. Baton Rouge; dwelling house and large brick sugar house.

No. 76. 933 acres, West Feliciana Parish, 20 miles from Bayou Sara village, near Laurel Hill. Can be divided into small farms to advantage.

No. 66. 949 acres, Tensas Parish, T. 11. Range 12. E; also, in Section 18, T. 10 Range 12 E. fine fresh high land: will make a good cotton plantation. lands are unimproved.

No. 64. 950 arpents, Parish of St. Tammany, on left bank of Tchefunta river unimproved timber land; a saw mill in good running order on the premises.

No. 51 D. 1000 acres, Rapides Parish, a fine tract of unimproved cotton land, partly in Grant Parish.

No. 45. 1000 arpents, Parish of Pointe Coupee, fronting 17 arpents on the Mississippi River, nearly opposite the town of Bayou Sara; 600 arpents cleared, balance fine timber; soil first quality, easy to cultivate, convenient to market, and easily drained.

No. 55. 1000 arpents, first class sugar plantation in the Parish of St. Mary; land high; beautiful yard, ornamented trees, orange trees; fine shrubbery; good machinery in old sugar house that was burned; runs back to high shell bank on Grand Lake, 8 miles above Franklin, fronts on Bayou Teche and Grand Lake. or Lake Chitemachus.

No. 62. 1000 arpents; a tract of fine timbered land in the Parish of St. Helena, on Amite river; fine location for saw mill, and for getting ship and railroad materials.

No. 21. 1015 acres good fresh land, in the Plaine's Settlement 10 miles from Baton Rouge; fine timber good stock range; can be divided into small farms to advantage.

No. 42. I120 acres, Parish of Natchitoches, 2 fronts on Red River; land good and productive, but needs improvement; near the town of Natchitoches and Grand Ecore landing: high dry and rich soil.

No. 41. 1035 acres, Madison Parish, consisting of 2 tracts; 795 on Joe's Bayou, 240 on Roundaway Bayou; fine cotton land; can be divided into small tracts; plenty of building and fencing materials.

No. 79. 1081 acres, Concordia Parish, north of Red River, on Tensas River; a valuable plantation.

No. 6. 1093 acres, No. 1 cotton land, east bank of Black river, Parish of Concordia; 1 dwelling, a stable, corn, cribs one gin in working order, cabins for 20 hands; other outhouses; land rich; good stock range.

No. 19. 1138 acres, good cotton land, E. Baton Rouge Parish; good soil, well located; can be divided into 3 or 4 good farms to advantage; 10 miles from Baton Rouge; plenty good timber, and good stock range.

No. 44. 1200 arpents, 40 arpents front on the Mississippi river; first class sugar plantation in the Parish of Plaquemines; large sugar mill, vacuum pans, and centrifugals.

No. 69. 1200 acres, unimproved tract in West Baton Rouge Parish; 500 acres cleared, balance in woodland; land is of superiar quality, suitable for cotton or sugar; 15 miles above Baton Rouge, 4 miles below Port Hudson; fine stock range; 50,000 bricks in a kiln on the place.

No. 84. 1,200 acres; an undivided fifth interest in this plantation, Bossier Parish.

No. 24. 1253 arpents: a well improved cotton plantation in Grant Parish, fronting on the left bank of Red River; one new cottage dwelling, outhouses, corn crib, stables, overseer's house, 12 cabins, store, etc. fine timber; place is in good running order.

No. 47. 1274 arpents, Parish Pointe Coupee, fronting 18 arpents on Mississippi river; fine location for a colony; land very productive; large amount of timber on it; good for cotton, sug ar, corn, etc.

No. 25. A. 1280 acres, fine cotton plantation in Grant Parish, frolnting on Red River, E. bank; 250 acres cleared and in a high state of cutivation; 10 good cabins and a small dwelling house.

No. 78. 1280 acres, St. Tammany Parish, on Thchefuncta river; good comfortable improvement, and other advantages.

No. 90. 1345 acres; one-fourth undivided interest in plantation formely belonging to Moses and Rachel Graves, Madison Parish.

No. 71. 1360 arpents, Parish West Baton Rouge; a first class, well improved and completely fitted out sugar plantation; 500 arpents fenced; good dwelling and cabins, large stable, large sugar house, slate roof; directly opposite Baton Rouge.

No. 56. 1380 arpents; a fine sugar plantation in the Parish of St. Mary, fronting on Bayou Teche, 8 miles above the town of Frank lin; large dwelling, beautiful yard, orange and ornamental trees; remains of old sugar house; plantation runs back to shell road on Grand Lake.

No. 52. 1600 acres; 400 acres woodland, balance cleared and fit for cultivation, 2 miles from Opelousas, Parish of St. Landry; healthy location; good society; splendid location for a colony; good for cotton, corn, sugar cane, fruits, sheep or stock raising. Can be divided into small lots to advantage.

No. 60. 1440 arpents, Parish St. James, on west bank Mississippi river; sugar and rice plantation; good dwelling: 20 double cabins, stable, large brick sugar house etc.

No. 5. 1440 acres: cotton plantation, Parish of Concordia; an improved place; has always made good average crops of cotton and corn rich land.

No. 35. 1480 acres, Parish of Iberville, on Bayou GrossTete; neat cottage dwelling and outhouses; good sugar and cotton lands; plenty fine timber; and other conveniences.

No. 51. 1500 arpents: 400 acres under fence, 600 more opened and ready for cultivation; property substantially and well improved; an excellent cotton plantation; fine timber; cane brake and wild pea for stock.

No. 59. C. 1520 arpents, Parish St. Bernard: a sugar plantation known as "Magnolia," fronting on both sides of Terre-aux-Bœuf; neat dwelling; 16 new cabins, a large sugar house in good order.

No. 59. 1600 arpents, Parish St. Bernard; fronts 20 arpents on each side of Bayou Terre-aux-Bœuf; land very rich, fine sugar and orange land; fine stock range.

No. 25. 1,625 arpents; fine cotton plantation in Grant Parish, on Red River, 5 miles below Colfax; dwelling, overseer's house, 15 cabins, stable, fine gin; fine, rich productive soil; land high and never fails to make good crops.

No. 28. 1,630 arpents; first-class, Parish of Iberia, fronting 20 arpents on Bayou Teche, 4 miles below Iberia; one large dwelling, 15 double cabins; 400 acres under fence; formely a fine sugar estate.

No. 75. 1689 acres of hill, creek bottoms and beach flatlands, Parish of West Feliciana, 4 miles from Bayou Sara, on the Port Hudson road; good for cotton; good location for a colony; can be divided to great advantage.

No. 3. 1695 acres, sugar and cotton plantation, Avoyelles Parish, on left bank of Bayou Bœuf; superior land; 1 good dwelling and outhouses; good cabins for 30 to 40 hands; sugar house; 150 acres cane.

No. 57. 1800 acres, Parish St. Bernard, 22 miles below New Orleans 700 acres in cultivation; a first class sugar plantation; well fitted up; good sugar house large sugar mill and machinery; place well ditched and in good order.

No. 38. 1840 arpents, a sugar and rice plantation, Parish of Jefferson, 1 mile above Kennerville, 12 miles by N. O. St. L. & C. Railroad from the city; 600 arpents under fence; one-half the place can be cultivated in sugar, the other half in rice, at the same time; large two-story dwelling, 8 rooms; excellent outhouses, 15 double cabins; very large brick sugar house, Nile's engine and centrifugals; 50 fine orange trees; valuable advantages, and a valuable plantation.

No. 37. 1842 arpents; a fine sugar plantation in the Parish of Jefferson; 12 miles from New Orleans, west bank of Mississippi river on Barataria Ridge; 800 acres under fence, balance fine cypress timber and good sugar wood; dwelling and outhouses complete, with oranges and fig trees surrounding the house, 15 cabins, corn house and stable, sugar house, with walking beam engine, in order; one set of open kettles; coolers, tankage, etc., in good order; a first-class stock, so near the city for marketing.

From Hon. Chas. Clinton, in regard to the 30-foot Wind Mill used for draining his Orange Grove.

STATE OF LOUISIANA, AUDITOR'S OFFICE, NEW ORLEANS, JAN. 29, 1876.

U. S. Wind Engine & Pump Co., Batavia, Ill.

Gentlemen:—Your favor of 24th inst. at hand. I received the box of weights you sent, but have not yet put them on the mill.

I am going down to the orange grove next week, and will look into the matter. My impression is that the mill does all the work well that we require of it, without the additional power given by the use of weights, still it is a good thing to know, and if I find it at all necessary, I will put them on and test the matter.

I am more than ever satisfied that your 30-foot mill is just what will eventually be used in this State for drainage, both on sugar and rice plantations. It only wants to be tried thoroughly, and if it does not get out of order within three or four years, and is well advertised, to start on, in some of the city local papers, it will give you a great many orders.

My mill works the two 16-inch pumps admirably, with a very light wind, and should it continue as now, without needing to be constantly repared, I shall feel that I have *the* power for drainage suited to this country.

Very respectfully, CHAS. CLINTON.

From noted Railway Officials concerning the DURABILITY of our MILLS.

BURLINGTON, CEDAR RAPIDS & MINN. R, R. CO.,
CEDAR RAPIDS, IOWA, APRIL 2, 1874.

John E. Thomas, Esq.,

Dear Sir:—Replying to yours of April 1st, I will say that we feel that fuel and water are two of the most essential supplies for the successful operation of a railroad, and have, therefore, experimented considerably with steam and horse power, as well as with different kind of wind mills, and our experience proves, to our complete satisfaction, that the Halladay Wind Mill is the best in the market. In fact, I know of no other which for durability, perfection of working, economy, and, I may truthfully add, mechanical beauty, that will begin to compare with this mill. We have sixteen of this manufacture on the line of our road and branches.

The entire expense of keeping these mills in repair, including six hand pumps, and, in fact, our whole water department, does not exceed one cent per mile per day. This includes all pump and wind mill repairs, also repairs on tanks and the putting in of new pumps. Respectfully yours,

W. H. PETTIBONE, *Asst. Gen. Supt.*

I fully endorse the above. WM. GREEN, *Vice-Pres. and Gen. Supt.*

BURLINGTON, CEDAR RAPIDS & MINN. R'Y CO.
Cedar Rapids, Iowa, Dec 16, 1876.

U. S. Wind Engine & Pump Co., Batavia. Ill.

Gentlemen:—Yours of 3d. inst. is received. In response, I would say: that upon the line of this railroad there are at present sixteen Halladay Wind Mills in operation—all in good repair, some of them having been erected nearly six years.

In reference to their durability, I will mention that something over three years ago, I erected at Burlington, on this road, a 22-foot Halladay Wind Mill, with 5-inch double acting force pump attached. The mill was set 90 feet above the ground. Some time after its erection a hurricane passed over that place; tracing its coarse by uprooting trees, and levelling, with other buildings, "Pond's Egg and Butter Packing Establishment," a brick building' killing and burying the proprietor, Mr. Pond, in the ruins The wind mill passed through that awful storm, as it were defiantly, and, if I may be allowed the expression, proudly proceeded to its work at the end of the blow; without having been damaged to the extent of one cent. This is an example of my uniform experience with the Hallady Wind Mill.

If the mill is properly erected, oiled about twice a week, and then left to take care of itself, nothing short of a tornado will disturb it.

Steam; so far as I have had experience; is 500 per cent, more expensive than wind power for pumping. You will understand this when I tell you that the expense for the labor of keeping in repair the wind mills and pumps of the entire water department of this railway, amounting to 400 miles of road, has never yet exceeded $4,50 per day, or 1⅛ cents per mile per day; and our water supply has always been abundant and satisfactory. JAMES CANTELO, *Master Builder.*

MORGAN LINE U. S. Mail Steamers.

The following low pressure iron Steamships form these lines:

AGNES,	CLINTON,	CITY OF NORFOLK,	AUSTIN,
W. G. HEWES,	HARLAN,	I. C. HARRIS,	GUSSIE,
WHITNEY,	HUTCHINSON,	JOSEPHINE,	MORGAN,
ST. MARY,			MARY.

Plying from Morgan City, in connection with Morgan's Louisiana and Texas R. R

FOR INDIANOLA VIA GALVESTON:

The following iron Steamers will leave Morgan City, as follows:

MORGAN .... On WEDNESDAYS.
CLINTON .... On FRIDAYS.
HUTCHINSON .... On SUNDAYS.

Freight for Indianola received. Connecting at Indianola with Gulf, Western Texas and Pacific Railway. Through bills of lading signed to San Antonio and intermediate points via Indianola and Gulf, Western Texas and Pacific Railway.

Through bills lading signed to Houston, and to all points beyond, on the Houston and Texas Central Railroad, and to Cuero, and to all intermediate points via Gulf, Western Texas and Pacific Railroad, subject to tariffs of respective lines.

Freight charged as per new card rates. Freight for Dallas and points beyond on the Texas Central Railroad, and to Troupe and points beyond on the International and Great Northern Railroad, taken at greatly reduced rates.

FOR CORPUS CHRISTI, FULTON and ST. MARY'S, via ROCKPORT—The iron steamer MARY, will leave Morgan City on ———.

Lighterage to Corpus Christi—if any—at risk and expense of consignees.

Freight for St. Mary's and Fulton landed at Rockport.

FOR BRAZOS SANTIAGO DIRECT—The iron Steamer I. C. HARRIS, will leave Morgan City MONDAY, Feb. 21.

Freight for all the above ports received at the Depot Morgan's Louisiana and Texas Railroad, foot of Lafayette street, daily, until 5 P. M.

Lighterage at Brazos Santiago at risk and expense of consignee. Freight payable in gold to Brazos Santiago.

Freight received for and bills lading signed only to Brownsville via Rio Grande Railroad, as per new tariff, less 8 per cent., until further notice.

Passenger and Mail Route.

FOR GALVESTON.

Passengers take Railroad Ferryboat, foot of St. Ann street, at 7½ A. M., reaching Morgan City at 11:15 A. M., there connecting with steamers—

HUTCHINSON .... On SUNDAYS,
JOSEPHINE .... On MONDAYS,
WHITNEY .... On TUESDAYS,
MORGAN .... On WEDNESDAYS,
JOSEPHINE .... On THURSDAYS,
CLINTON .... On FRIDAYS,
WHITNEY .... On SATURDAYS.

Cabin Passage .... $12 00 Deck .... $ 6 00

First class fare to Shreveport, $25—time 45 hours. Excursion Tickets for the round trip to Galveston and return, good for thirty days, issued at $21.

Tickets and staterooms secured until 5 P. M. daily—Sundays excepted—at Agent's Office, or at the Ferry Landing, on morning of departure. Through Tickets will be issued from this Office to the principal points on the Houston, Texas Central Railroad, on the International and Great Northern Railroad, Texas and Pacific Railroad, and Gulf, Western Texas and Pacific Railway, and Stage connections.

C. A. WHITNEY & CO., Agents,
Corner Magazine and Natchez Streets.

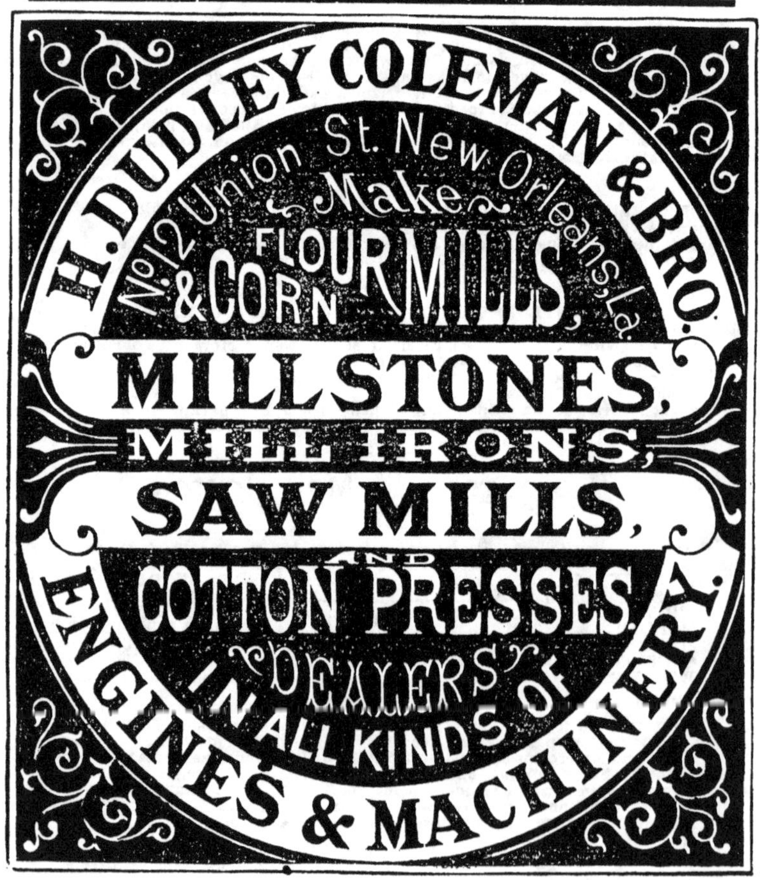

H. DUDLEY COLEMAN & BRO.
No. 12 Union St. New Orleans, La.
Make
FLOUR & CORN MILLS,
MILLSTONES,
MILL IRONS,
SAW MILLS,
AND
COTTON PRESSES.
DEALERS IN ALL KINDS OF
ENGINES & MACHINERY.

Chemical and Blood Fertilizers.

VILLE
Fertilizers!

MANUFACTURED BY THE

CHEMICAL

—AND—

Blood Fertilizing WORKS!

—OFFICE:—

37 NORTH PETERS STREET,

Between Bienville and Customhouse Streets,

FERTILIZER No. 1.—FOR SUGAR CANE.

FERTILIZER No. 2.—FOR COTTON.

FERTILIZER FOR ORANGE AND FRUIT TREES.

FERTILIZERS.

For Sugar Cane	$60 00
For Cotton	45 00
For Orange and Fruit Trees	60 00
Pure Super Phosphate	65 00
Pure Ground Bones	45 00

Per ton, in sacks of 200 lbs.

Chemicals Always on Hand:

NITRATE OF POTASH,
NITRATE OF SODA,
MURIATE OF POTASH,
LAND PLASTER,
SULPHURIC ACID,
SULPHATE OF AMMONIA.

Fertilizers compounded on Orders for all Crops.

☞ This branch a specialty.

—OFFICE:—

37.........NORTH PETERS STREET.........37

Between Bienville and Customhouse.

Chemical and Blood Fertilizers.

Chemical and Blood Fertilizers.

Chemical and Blood Fertilizers.

INDEX.

Area of Louisiana....................14...15
Arable Alluvial Lands..........15, 22, 24, 25
Alluvial Lands, Wooded Swamps...... 15, 19
Alluvial Parishes..........................13
Abandoned Lands...........................36
Attakapas and St. Landry 38
Au Large Prairie...........................59
Agricultural Products 92

Bays.................................21...15
Bundick's Creek Settlement 81
Bayou Teche ....58...70
Bones and Baskets (salt mine) 64
Bill of Fare of a Louisiana Farmer 123
Bee Culture 189
Breaux Bridge 75
Bean Bassin.................................79
Bayou Cypremort 101
Bluff Lands......................15, 26, 27

China trees—fuel...........................133
Coast Marshes...............................19
Coast Bayous................................21
Coast Line..................................21
Cotton Lands—Best in the World............34
Cheap Sugar House 60
Creole Industry 60
Crops...................................77, 84
Cote Gelée Hills............................80
Climate.................................96, 97
Cote Blanche Island 102
Cattle in Cameron 114
Crops in Cameron...........................117
Cheap Sugar Estates........................119
Crops, per Hand and per Acre 122
Chief Wants of Louisiana...................185
Cattle and Sheep...........................187
Cotton Seed Analyzed.......................198
Crops of Louisiana compared with Western crops 122

Dairying in Cameron........................114
Domestic Fowls—prices.......................55
Duck Hunters 235

Exhaustless Natural Wealth, ...35, 121, 129, 184
Exalted Destiny of Louisiana 185

Fruits and Vegetables 41, 94
Farm Crops, prices 55
Fine Estates................................72
Facts—Wonderful Crops 121
Fruit Culture 84, 88, 140
Fruits from Gulf Coast... 150

Guinea Grass in Pinelands..................219
Good Uplands 15, 32
Geological View.............................24
Grande Côte Island 61
Grande Pointe 76
Game, Fishes, Oysters..............76, 85, 89
Grand Chenier Island........................90
Gardens 93
Geographical Names—Pronunciation 104
General Elevation (St. Mary) 92
Grasses and Forage, Plants 202

Hays, John 65
Hilly Lands 14
Homes on the Lakes 20
Hog Wallow Lands 31
Health of Vermilion Parish..................83
Hedges 100
Health...............................96, 111, 170
Health of White Laborers 159

Islands in the Marsh....................20, 28
Immigration Needed 37
Illinois Farmer—His Views...................42
Illinois Editor—His Views 42
Iberia Parish 57
In the Wilderness...........................74
Indiana Farmer...His Views 188
Innate Value of Louisiana Soil 201

Jetties 243

Level Lands.................................14
Longevity 56, 71
Lake Peigneur 66
Lake Tasse 68
Lake Martin.................................75
Lake Charles...............................111

Mississippi River...Distances 229
Morgan Railroad 43
Morgan Steamships 43
Minerals..............................190, 112
Moss...Spanish.......................129, 238
Most Valuable Timber 131
Mortality...White and Colored 180
Missouri Planter in Louisiana..............183
Mineral Springs 187
Molasses Analyzed 194
Medical Flora of Louisiana 200
Mortality..........................247 to 251
N. E. Louisiana...Over the Lake 230
New Orleans...Population 228
Natural Divisions 17
North Louisiana Villages 30

This Page is Blank

Materials were not available when scanned.

INDEX.

Area of Louisiana....................14...15
Arable Alluvial Lands..............15, 22, 24, 25
Alluvial Lands, Wooded Swamps...... 15, 19
Alluvial Parishes..............................13
Abandoned Lands..............................36
Attakapas and St. Landry 38
Au Large Prairie................................59
Agricultural Products 92

Bays...21...15
Bundick's Creek Settlement 31
Bayou Teche .... 58...70
Bones and Baskets (salt mine) 64
Bill of Fare of a Louisiana Farmer 123
Bee Culture 189
Breaux Bridge 75
Bean Bassin...79
Bayou Cypremort 101
Bluff Lands............................15, 26, 27

China trees—fuel................................133
Coast Marshes.......................................19
Coast Bayous...21
Coast Line...21
Cotton Lands—Best in the World............34
Cheap Sugar House 60
Creole Industry 60
Crops..77, 84
Cote Gelée Hills.....................................80
Climate..96, 97
Côte Blanche Island 102
Cattle in Cameron 114
Crops in Cameron..................................117
Cheap Sugar Estates..............................119
Crops, per Hand and per Acre 122
Chief Wants of Louisiana......................185
Cattle and Sheep...................................187
Cotton Seed Analyzed...........................198
Crops of Louisiana compared with Western crops 122

Dairying in Cameron.............................114
Domestic Fowls—prices...........................55
Duck Hunters 235

Exhaustless Natural Wealth, ...35, 121, 129, 184
Exalted Destiny of Louisiana 185

Fruits and Vegetables 41, 94
Farm Crops, prices 55
Fine Estates...72
Facts—Wonderful Crops 121
Fruit Culture 84, 88, 140
Fruits from Gulf Coast... 150

Guinea Grass in Pinelands....................219
Good Uplands 15, 32
Geological View.......................................24
Grande Côte Island 61
Grande Pointe 76
Game, Fishes, Oysters.................76, 85, 89
Grand Chenier Island...............................90
Gardens 93
Geographical Names—Pronunciation 104
General Elevation (St. Mary) 92
Grasses and Forage, Plants 202

Hays, John 65
Hilly Lands 14
Homes on the Lakes 20
Hog Wallow Lands 31
Health of Vermilion Parish........................83
Hedges 100
Health 96, 111, 170
Health of White Laborers 159

Islands in the Marsh..........................20, 28
Immigralion Needed 37
Illinois Farmer—His Views........................42
Illinois Editor—His Views 42
Iberia Parish 57
In the Wilderness..74
Indiana Farmer...His Views 188
Innate Value of Louisiana Soil 201

Jetties 243

Level Lands..14
Longevity 56, 71
Lake Peigneur 66
Lake Tasse 68
Lake Martin...75
Lake Charles..111

Mississippi River...Distances 229
Morgan Railroad 43
Morgan Steamships 43
Minerals 190, 112
Moss...Spanish..................................129, 288
Most Valuable Timber 131
Mortality...White and Colored 180
Missouri Planter in Louisiana...................183
Mineral Springs 187
Molasses Analyzed 194
Medical Flora of Louisiana 200
Mortality.......................................247 to 251
N. E. Louisiana...Over the Lake 230
New Orleans...Population 228
Natural Divisions 17
North Louisiana Villages 30

Oranges...Great Demand....149
Orange Island....67
Orange Crop....94, 112, 117
Overflows....101
Oak Bark and Tanneries....54
Oak Staves....132

Pine Hills....15, 30, 32, 216, 217
Prairies....15, 22
Pine Flats....15, 29
Panoramic View....18
Pine Lumber...Vast Supplies....31
Products of Good Uplands....35
Prairie Herds....44
Price of Lands....49, 71, 79, 88, 188
Pierre Broussard....61
Pisgah View....64
Poultry....72
Parish of St. Martin....68
Parish of Lafayette....78, 79, 83
Prices...Crops, Cattle, Poultry....84
Parish of Vermilion....86
Poor Immigrant's Parish....90
Parish of St. Mary....91
Population....97
Profits of Small Sugar Farms....99
Plantations and Farms, etc....98, 118, 119
Poet's Views...Evangeline....105
Parish of Calcasieu....112
Parish of Cameron....113
Prof. Hilgard's Opinion....41
Pitch, Tar and Turpentine....132
Petroleum....135
Profits of White Labor....158, 184
Piny Woods....32
Petite Anse Island....63
Parish of St. Landry....43
Pear Culture on Bayou Teche....147
Pine Straw...Analysis....218
Pineland Farmers...What they may do...221
Population...Table....227
Public Lands....243

Rice Crop....93, 135; 224
Rich Lands in Louisiana....128
Railroads....227
Rainfall....247

Sugar Crop...Table....224
Soil of Lafayette Parish....78
Soil and Scenery....87
Sugar Crop of St. Mary....103
Saw Mills....111
Sheep, Hogs, Poultry....115
Silk Culture in Louisiana....124
Soil Resources....123
Sumac....126
Sulphur Mine....135
Scuppernong Grapes....145
Statement of New Orleans Physicians....179
Summer Heat....185
Salt Mine....28, 64, 65, 134
Scenery in North Louisiana....34
St. Tammany Parish....186
Southern Onions....201

Timbered Bottoms....45
Teche Forests....69
Tobacco....94
Tea Culture in Louisiana....126
Temperature of Climate....245

Value of Cotton Seed Cake....198
Vermilion River....81
View on Horseback and on Foot....18

What a Poor Man Can Do....162
White Men in Cane, Cotton and Rice Fields....155, 156, 157, 160
Wealth of the Forests of Louisiana....130
Water Surface....15, 20
Wealth and Refinement....18
White Labor....51, 52, 53, 83, 154
Wild Fowls in Cameron....116
What Two Men Did....163
What White Boys Can Do....165
White Tenants in Iberville....164
Water Power....187

Yield of Sugar, Cotton, Corn....49
Yellow Fever....177, 181
Yellow Fever...Hedging In....221

www.ingramcontent.com/pod-product-compliance
Lightning Source LLC
LaVergne TN
LVHW010219110826
845151LV00004B/1143
* 9 7 8 1 4 2 5 5 2 6 0 4 7 *